AVIATION GUIDE

I.A.R.80 & I.A.R.81

by Radu Brînzan

Aviation Guide No.3
I.A.R.80 & I.A.R.81

First published in 2011 by SAM Limited, under licence from SAM Publications
Media House, 21 Kingsway, Bedford, MK42 9BJ, United Kingdom

© 2011 SAM Publications
© Radu Brînzan – Text
© Radu Brînzan – Colour Artwork
© Radu Brînzan – Scale Plans/Scrap Views

ISBN 978-1-906959-19-7

Series Editor – Andy Evans
Designed by Jonathan Phillips
Typeset by SAM Publications, Media House, 21 Kingsway, Bedford, MK42 9BJ,
United Kingdom
Printed and bound in the United Kingdom by acorn Web Offset Limited

About the Author

Radu Nicolae Brînzan was born in Galaţi, Romania in 1971. He moved to Ireland in 1995 for third-level studies. He works as a free-lance translator on behalf of a number of Irish Government agencies. He is a passionate scale-model builder and the holder of a variety of medals and awards in international scale-model competitions. As a scale-model builder, he is also the author of a number of model-related articles published in Scale Aviation Modeller International and on the Internet. He is also a passionate follower of Romanian aviation he has translated a number of Romanian aviation books into English and provided help as a technical adviser on Romanian related books and articles. He has also produced a number of resin patterns and photo-etched accessories, as well as decal artwork for various model companies and is currently the owner and product developer of RB Productions. He and his wife Elaine live in West Cork.

Glossary of Terms

- Aeronautica Civilă - Civil Aviation, part of Aeronautica Regală Română
- Aeronautica Militară - Military Aviation, part of Aeronautica Regală Română
- Anvelope Banloc - Banloc Tyres, manufacturer of rubber tyres, the Romanian branch of the US manufacturer BF Goodrich
- Anvelope Floreşti - Floreşti Tyres, manufacturer of rubber tyres, the Romanian branch of the British manufacturer Dunlop
- ARR - Acronym of Aeronautica Regală Română = Romanian Royal Aviation
- Asalt = Attack, as in 'Aviaţie Asalt' = Ground attack aviation
- ASAM - Acronym of Asociaţia Stabilimentelor Aeronauticii Militare = Association of Military Aviation Establishments. This association ran a number of workshops that serviced, overhauled and repaired aircraft
- BoPi – abbreviation of 'Bombardier in Picaj' = Dive Bomber
- Centrul de Instrucţie al Aeronauticii = Aviation Training Centre
- Comandamentul Aeronauticii Militare = Military Aviation Command, the renamed Gruparea Aeriană de Luptă after June 1946
- Comandamentul Aviaţiei = Aviation Command, the renamed Comandamentul Aeronauticii Militare after February 1949
- Comisia Militară = Military Commission, a military inspection body
- Compania Terestră si Hidro = Ground and Sea Equipment Company
- Corpul Aerian Român = Romanian Air Corps abbreviated to CAR, part of Gruparea Aeriană de Luptă
- Corpul 1 Aerian Român = First Romanian Air Corps, the name of Romanian fighter aviation after September 1944
- Direcţia Tehnică = Technical Department
- Elektron – an alloy of magnesium and other light non-ferrous metals, used extensively for die-cast aviation parts
- Escadrila de Experienţe = Test Squadron based on Pipera Airfield created in September 1940 for the purpose of evaluating the I.A.R.80
- Escadrila de Alarmă I.A.R. Braşov = Scramble Squadron of I.A.R. Braşov, the factory defence unit
- Fabrica de Celuloză Zărneşti = Zărneşti Cellulose Factory, paper and cardboard manufacturer located in Zărneşti, near Braşov
- Forţele Aeriene Militare = Military Air Force, the official name of the Romanian military aviation after 1 April 1950
- Gruparea Aeriană de Luptă = Air Combat Grouping, the expeditionary detachment of Aeronautica Militară
- I.A.R. - Acronym of Industria Aeronautică Română = Romanian Aviation Industry
- Ing.Dipl. - Abbreviation of Inginer Diplomat = Bachelor of Engineering
- Inspectoratul General al Aeronauticii = General Inspectorate of Aeronautics
- Întreprinderea Metalurgică de Stat - State Metallurgical Enterprise, the name of I.A.R. Braşov after June 1947
- I.O.R. - Acronym of Industria Optică Română = Romanian Optical Industry
- MAM - Acronym of Ministerul Aerului şi Marinei = Air and Marine Ministry, until 1948
- MFA - Acronym of Ministerul Forţelor Armate - Ministry of Armed Forces, after 1948
- Ministerul de Război = War Ministry
- Picaj = Dive
- PREROM - manufacturer of aviation instruments based in Săcele, near Braşov, the Romanian branch of the Czechoslovak company PREMA
- P.Z.L. - Acronym of Państwowe Zakłady Lotnicze - Polish State Aviation Works
- Regia Autonomă I.A.R. Braşov - I.A.R. Braşov Independent Board, the name of I.A.R. Braşov after nationalisation in 1938
- S.E.M.A.T = Manufacturer of hydraulic/pneumatic devices
- Serviciul Tehnic = Technical Service
- SMA - Acronym of Statul Major al Aerului = Air Chief of Staff
- SMP - Acronym of Serviciul Militar de Propagandă = Military propaganda Service, the source of many photographs included in this book
- Sovrom Tractor - Soviet-Romanian Tractor Co-operative, the name of I.A.R. Braşov after 1946
- SSA - Acronym of Subsecretariatul de Stat al Aerului = State Junior Air Ministry, a branch of MAM
- Şcoala de Vânătoare = Fighter Aviation School, abbreviated to Şc.Vt.
- Şcoala Militară de Aviaţie = Military Aviation School. This was abbreviated to Şc.Mil.Av.
- Uzina de Tractoare Braşov = Braşov Tractor Factory, the name of Întreprinderea Metalurgică de Stat after 1948
- Vânătoare - When translated literally, it means 'to hunt' and is used in association with a number of specialties of the Romanian armed forces, such as 'tank hunter' or 'mountain hunter'. When used in an aviation context, it means 'Fighter Aviation', often abbreviated to Vt.
- VDM - Acronym of Vereinigte Deutsche Metallwerke G.m.b.H. = Allied German Metalworks Ltd., propeller manufacturer

Bibliography

- "20mm Flugzeug-Mehrfachkanone 'Oerlikon' Beschreibung, Bedienungs und Einbauanleitung", Werkzeugmaschinenfabrik Oerlikon, Zürich-Oerlikon / Schweiz
- "222 Best Romanian Royal Air Force Pictures 1941 - 1945", Cristian Crăciunoiu & Ion Dobran, Editura Modelism, Bucharest, 2008
- "Armata Română 1941 - 1945" Cornel I. Scafeş, et. al, Editura Rai, Bucharest 1996
- "Aviatia Română în Prima Zi de Război, 22 Iunie 1941", Dan Antoniu and George Cicoş, Editura Phoebus, Galaţi, 2007
- "Aviaţia Romană în Timpul Celui de-al Doilea Război Mondial", Medin Robănescu & Teodor Liviu Moroşanu, Editura AeroMagazin, Bucharest, 2005
- "Beschreibung und Bedienungsvorschrift der Fl.-Bordfunkanlage, Baumuster Isl.2a (FuG. VII)" Werksschrift von Telefunken, 1938
- "Bf 109F 1, 2, 3 Beschreibung, Bedienungs-, Wartungs- u. Prüfvorschrift der Bordfunkanlage", Der Reichsminister der Luftfahrt und Oberfehlshaber der Luftwaffe, Berlin, November 1940
- "British Aircraft Armament, Volume 2: RAF Guns and Gunsights From 1914 to the Present Day", R. Wallace Clarke, Patrick Stephens Limited, Sparkford, 1994
- "British Aviation Colours of World War Two, The Official Camouflage, Colours and Markings of RAF Aircraft, 1939-45", John Tanner et al, RAF Museum, Hendon, 1986
- "Browning Machine Gun .30 Calibre Air Cooled", Fabrique Nationale D'Armes de Guerre, S.A., Herstal Belgique
- "Catalog cu Piese de Schimb, Avionul I.A.R.81", Subsecretatiatul de Stat al Aerului, Direcţia Tehnică Aeronautică, Regia Autonomă Industria Aeronautică Română Braşov
- "D.(Luft) T.2190 A-7 bis A-9, "FW190 A-7 bis A-9 Flugzeug Hanbuch, Teil 8 C Sonderwaffeanlage: 21 cm BR", September 1944
- "D.(Luft) T6151 "MG151 und MG151/20, 15mm-Flugzeugmaschinegewehr 151 und 20mm-Flugzeugmaschinegewehr 151/20", Mauser-Werke Oberndorf, 1942
- "Die Leuchtpistole und ihr Gebrauch", Heinz Denckler, Heinz Denckler Verlag, Berlin, 1942
- "Entwurf einer Beschreibung und Bedienung und Wartungsvorsicht für den Höhenatmer HLa732 und die Höhenatemmaste HM 5 un HM15", Berlin, 1938
- "Entwurf einer Beschreibung und Bedienungsvoschrift des MG-FF (2 cm Flugzeug MG-FF) Ausfuhrung A", Berlin, 1938
- "Ersatzteilliste [Truppenersatzteilliste] für die VDM-Verstelluftschraube, Ausgabe Oktober 1941", Vereinigte deutsche Metallverke A.G., Frankfurt (Main) - Heddernheim
- "Ersatzteilliste für das 2 - cm - Flugzeug - MG - FF, MG-FF M", March 1940
- "Flying Guns of World War II", Anthony G. Williams & Emmanuel Gustin, Airlife, Shrewsbury, 2003
- "From Barbarossa to Odessa" Volumes 1 and 2, Dénes Bernád, Dmitry Karlenko and Jean-Luois Roba, Ian Allan Publishing, Hersham, 2007 & 2008
- "German Aircraft Cockpits 1911 - 1970", Peter W. Cohausz, Schiffer Military History, Atglen PA, 2003
- "German Aircraft Interiors 1935-1945", Kenneth A. Merrick, Monogram Publications, Boylston, 1996
- "German Air-Dropped Weapons to 1945", Wolfgang Fleischer, Midland Publishing, Hersham, 2003
- "Hohen-Atmungs-Gerat 'Munerelle' Type Agm.40, Bedienungsvorschrift fur die Flugzeugbesatzungen", Schweitzerische Armee, Dienstakt nr. 202a d/f, Ausgabe 1.9.47
- "I.A.R.80", Antoniu & George Cicoş, Editions TMA, Paris, 2008
- "Luftwaffe Camouflage and Markings 1933 - 1945, Volume One", K. A. Merrick, Chevron Publishing Limited, Hersham, 2004
- "Luftwaffe Camouflage and Markings 1933 - 1945, Volume Two", K. A. Merrick, Chevron Publishing Limited, Hersham, 2005
- "Macchi MC 200 Saetta, pt.2a, Aviolobri Special 9" Maurizio di Terlizzi, Istituto Bibliografico Napoleone, Rome, 2004
- "Nomenclatorul Avionului I.A.R.80", I.A.R. Braşov, 1941, Filiala Arhivelor Statului Braşov, Fondul Nr. 522 - Fondul I.A.R., Dosar Nr.65,
- "Notiţa Tehnică a Avionului de Recunoaştere şi Bombardament I.A.R.37 cu Motor I.A.R.-K 14-IV C 32", Subsecretariatul de Stat al Aerului, Direcţia Tehnică Aeronautică, S. E. T. - Industria Naţională Aeronautică, 1943
- "Notiţa Tehnică a Avionului de Vânătoare I.A.R. 80 cu Motor I.A.R. 14 K IV. c. 32", Subsecretariatul de Stat al Aerului, Direcţia Construcţiilor Aeronautice, Regia Autonomă Industria Aeronautică Română, Braşov, 1941
- "Notiţa Tehnică a lansatorului de bombe pentru picaj a avionului I.A.R. 80" Industria Aeronautică Română, Braşov, 1942
- "Notiţa Tehnică Descriptivă a Avionului 'Savoia-Marchetti' S.79B", Ministerul Aerului şi Marinei, Aeronautica Militară,
- "Piese de Schimb, Avionul IAR-38", Industria Aeronautică Română Braşov, 1939
- "Ploiesti: Low Level Strike", Cornel Năstase, Dan Melinte, Răzvan Bujor, Modelism, Bucharest, 2007
- "Pr. Ver. Conv. Germania", Braşov, 1942, Filiala Arhivelor Statului Braşov, Fondul Nr. 522 - Fondul I.A.R., Dosar Nr. 93,
- "Proces - Verbal încheiat de Comisia Flotilei 2 Vânătoare pentru verificarea modului de comportare a bombardamentului în picaj a avionului IAR-80", Report issued by Flt.2Vt. to I.A.R. Braşov, 1941
- "PZL P.11, Cz.1" A. Glass, T. Kopanski, T. Markovski, AJ-Press, Gdynia, 1997
- "PZL P.11, Cz.2" A. Glass, T. Kopanski, T. Markovski, AJ-Press, Gdynia, 1997
- "PZL P.11c" Bartolomej Belcarz & Tomasz J. Kopanski, Mushroom Model Publications, Redbourn, 2003
- "PZL P.24", Andrzej Glass, Kagero Publications, Lublin, 2009
- "PZL P.24", Przemisław Skulski, ACE Publications, Wrocław, 2002
- "Romanian Aeronautical Constructions", Dan Antoniu & George Cicoş, self-published, Bucharest, 2007
- "Romanian Aeronautics in the Second World War", Cristian Crăciunoiu & Jean-Louis Roba, Editura Modelism, Bucharest, 2003
- "Rumanian Aces of World War 2", Dénes Bernád, Osprey Publications, Northants, 2003
- "Rumanian Air Force, the Prime Decade, 1938 - 1947", Dénes Bernád, Squadron/Signal Publications, Carrollton, 1999
- "The Bombardier's File", US War Department, AAF Form 24N, Approved 11-23-44
- "The Browning Machine Gun, Volume II, Rifle Caliber Brownings Abroad", Dolf L. Goldsmith, Collector Grade Publications, Ontario, 2006
- "The Official Monogram Painting Guide to German Aircraft 1935 - 1945", Kenneth A. Merrick & Thomas H. Hitchcock, Monogram Aviation Publications, Boylston, 1980
- "Third Axis - Fourth Ally, Romanian Armed Forces in the European War 1941 - 1945", Max Axworthy, Cornel Scafeş, Cristian Crăciunoiu, Arms and Armour Press, London 1995
- "Vânătorul I.A.R.80, Istoria unui erou necunoscut", Dan Antoniu & George Cicoş, Editura Modelism, Bucharest, 2000
- "Первый Як" [First Yak], S. Kuznetzov, Polygon, Moscow, 1995

Periodicals (in alphabetical order)

- Aero-Magazin, Bucharest No. 8, February 2003
- Air International, May 1990
- Aripi Româneşti, official publication of Subsecretariatul de Stat al Aerului, Bucharest 1942 to 1944
- Der Adler, Romanian language edition of the German magazine, Berlin: 1942 to 1944
- Magazin Aeronautic, official publication of Subsecretariatul de Stat al Aerului, Bucharest, 1942
- Modelism Magazine, Bucharest: 3/1984-(4); 4/1884; 3/1985-(8); 2/1989-(23); 3/1989-(24); 4/1989-(25); 3/1991
- Top Gun Special Magazine, Bucharest: February 2000; April 2000; June 2000

Contents

Acknowledgements

Thanks and appreciation to Dan Antoniu and Răzvan Bujor who generously shared their archives and vast knowledge, to George Cicos who, supplied the individual aircraft data and a large amount of historic information, and last but not least I must thank Dan Melinte who generously made available all his knowledge about this beautiful aircraft. Thanks also to my brother Florin George Brînzan for helping me with the many line drawings, and to my parents Emil and Domnica Brînzan for supplying me with what seems to be every single book and magazine on Romanian aviation published the last twenty years! Thank you Elaine for your patience!

Thanks are also due to:

ABC Collection – (Dan Antoniu, Răzvan Bujor and George Cicos collection), Arthur Bentley, Dénes Bernád Dan Cătălin Buzdugan, Philippe Couderchon of the Memorial Flight Association, Franck Devillers of Collection Ailes Anciennes le Bourget, Jose Fernandez, Christian Gloor of www.flightgear.ch, Günter Hütter, Dan Iloiu, Stanislav Zharkov, Viktor Kulikov, Jean-Francois Legendre, Kenneth Merrick, Jon Moran of www.browningMG.com, Teodor Liviu Moroanu for his help with the colour drawings, particularly squadron insignia and cartoons, Romanian Military Archives, Horia Stoica, Dave Wadman, Erwin Wiedmer of www.germancockpits.com and Anthony G. Williams

Introduction

The Most Famous Fighter You Never Heard Of

The I.A.R.80/81 was Romania's only home-designed fighter to see action in World War II, during which, against all sorts of odds, it faced and successfully defeated powerful foes packing a bigger punch. Fate made it so that this aircraft managed to get into fights with some of the major players of World War II, the Russian VVS, the US Air Force, the German Luftwaffe, as well as the Hungarian Légierő, while always fighting for the same air force, Aeronautica Regală Romănă, the only air force it ever flew for. Within the period of a little more than three years that the aircraft saw combat, pilots flying the I.A.R.80/81 claimed an estimated total of 539 confirmed and 90 unconfirmed victories against aircraft in the air and 168 confirmed victories against aircraft on the ground. However, the price paid was 274 I.A.R.80/81 lost to combat, anti-aircraft fire and accidents. Only 450 I.A.R.80/81 was ever made. Between 1941 and 1945, Comandamentul Aeronauticii Militare recorded around 1200 confirmed aerial victories by Romanian fighters. Placing the above-mentioned tallies in this context makes the I.A.R.80/81 one of the most effective fighter aircraft used by Aeronautica Regală Romănă in World War II. Considering that Aeronautica Regală Română was the third largest air force on the Eastern Front after the Russian VVS and the German Luftwaffe, one would expect this aircraft to be world-famous. Why is it not? Well, the reasons may be many, but the single-most important one is that it was just one more victim of historically turbulent times that were intentionally brushed under the carpet and wilfully forgotten. The I.A.R.80/81 happened to be developed, used and disposed of during a decade when the fate of Romania was changed by dictatorship, territorial loss, war, defeat and occupation. During this period, the administration of the country swung from a Constitutional Monarchy to a People's Republic, the economy collapsed from a wealthy oil exporter and breadbasket to an economic basket case while the political system sampled the entire political spectrum with a radical shift from extreme right to extreme left, all of which brought their own forms of tyranny and hardship. After all of this, the country was smothered behind the Iron Curtain, in the grip of a cruel, incompetent and failing political system that became increasingly paranoid and mired in self-imposed isolation as the years went by.

Before World War II and for almost two years after the beginning of the conflict, Romania was neutral. However, this did not stop its neighbours from claiming and seizing large parts of its territory. These losses led to fierce nationalism and the rise of extreme right-wing fascist movements, which gradually secured a foothold, seized power and eventually led Romania towards joining the Axis and then going to war with the Soviet Union and the Allies. Just like any other cog in the Romanian war machine, the I.A.R.80/81 did its best and fought heroically against the Soviet forces during this war, but the price paid for this heroism was high. When the war ended and Romania found itself in the Soviet sphere of influence, with Soviet troops based in Romania for the two decades that followed the war, many of the aviators, engineers, mechanics or any people that were in any way involved in the war with the Soviet Union ended up suffering all kinds of trouble for it. Soldiers, most of whom truly had no choice but fight as ordered, had to pay dearly and feel ashamed for the simple misfortune of being young enough to be conscripted. In the immediate post-war years, any remaining career officers and soldiers who fought in the war against the Soviet Union were purged from the armed forces reorganised under Soviet supervision. With a new constitution denouncing any association with "fascism", it quickly became not only imprudent but also almost illegal to speak about and confess to wartime co-operation with Fascist Germany. Like any other Communist countries, Romania quickly put in place its own pervasive and oppressive secret police, the 'Securitate', and the fear of "informants" led people to be careful about what they spoke and to whom, but most said nothing to anyone. History that was still fresh in people's minds was rewritten or distorted, those who could correct it were unable to do so and the war on the Eastern Front was treated as if it did not happen at all. For example, when studying school history in Communist Romania, the campaign against the U.S.S.R. was not even alluded to (that period was summarily and confusingly described as the 'Fascist occupation of Romania') and the 'War' simply 'began' for Romania on 23 August 1944 when Romania joined the Soviet Union in the fight against Germany following a "People's Insurrection" led by the Communist Party. This 'mentionable period' between August 1944 and VE Day was routinely and conveniently described as "the struggle to liberate the motherland from the Hitlerist yoke, under the leadership of the Communist Party", not a single word of which was an accurate description of the events or the actors involved. The real tragedy is that this oppressive climate and denial of historic truth was at its peak for the two decades after the war, while there was still a Soviet presence in the country, and thus a golden opportunity to gather information while memories were still fresh was lost for ever.

However, the I.A.R.80/81 was not altogether forgotten in Romania and it was mentioned occasionally in magazines and books, but again only in the context of what the Communists called the 'Anti-Hitlerist War', respectively the last eight months of the war. Sadly, by that stage of the conflict, the I.A.R.80/81 was almost obsolete and there were few stories of bravery and success involving it that could be told. To compound these problems, Romania was stuck behind the Iron Curtain until 1990 and as a result, the Western historians' access was mostly limited to information about Aeronautica Regală Română available from German archives, which was often scant or incomplete. As a result of this wilful neglect from within and inaccessibility from beyond the borders of Romania, the I.A.R.80/81 gradually found itself relegated to a bit-part in the history of aviation.

As already mentioned, the spark of interest for the I.A.R.80/81 was kept alight by a number of enthusiasts and historians who never forgot this charismatic aircraft. The book

I.A.R.80–A No. 60 of Şcoala de Vânătoare Flt.3Vt. photographed on a snowy Galaţi airfield in early 1943. Note the towing device used to lift the aircraft by the tail skid and wheel it around the airfield.

Photo courtesy of ABC Collection

'Construcţii Aeronautice Româneşti' [Romanian Aeronautic Constructions] written by Ovidiu Ionescu and published in 1970 contained a brief history of the aircraft and a basic set of drawings. Another major step was taken by the Romanian magazine Tehnium that published a basic set of drawings in 1975, which was a great start, but sadly, they contained a number of errors. In 1976, 'Air International Magazine' published a set of drawings based on these Tehnium drawings, which served for many years as inspiration for many inaccurate scale models. Later, Tehnium's sister/off-shoot magazine 'Modelism' published a set of drawings and illustrations from the technical manuals in 1984. 'Modelism' continued this trend in subsequent issues by publishing some photographs and some striking artwork. Later, in 1989/1990, 'Modelism' reached a major landmark when they published the very detailed drawings by Dan Iloiu, which was also the precise moment when my interest in this beautiful aircraft began in earnest. Fortunately, that coincided with the anti-communist revolution in Romania and as a result of this new-found freedom, information became much more readily available. Ten years later, the book 'I.A.R.80, Istoria unui vânător necunoscut' [I.A.R.80, The History of an Unknown Fighter] by Dan Antoniu and George Cicoş first published in Romanian by 'Modelism' in 2000 and later in French by Editions TMA in 2008 completed the task of bringing this beautiful aircraft back to memory. Dan Antoniu's research remains the main source of information about the I.A.R.80/81 and his vast knowledge and generous advice are firmly placed at the foundation of this book. In fact, the present book should be read in conjunction with Dan Antoniu's aforementioned work that contains a wealth of historical information, an almost day-by-day diary of I.A.R.80/81 service and very moving stories of heroism and sacrifice.

Although aircraft of many other nations were given evocative monikers such as Würger, Spitfire, Tomahawk, Chayka, Mustang, Typhoon, Hayabusa or Orel, the I.A.R. 80/81 was always called just that. In Romania, it is simply known as the 'Yee-Ah-Reh', which is the way Romanians pronounce I.A.R. This book is the culmination of two decades of increasingly intense study and research of this beautiful aircraft. The line drawings included in this book were created over a period of nine months and are based on dimensions obtained from original documents, study of many photographs and the existing drawings by Dan Iloiu. The purpose of this book is to describe the aircraft in as much technical detail as possible, concentrating on all aspects such as structures, systems, and weapons. My intention was to put together a 'walk-around' - style book, but in the absence of any preserved airframes, I had to rely on images from the technical manuals and part lists. The difference between the 80 and 81 designations simply refers to the purpose of the aircraft whereby the I.A.R.80 was a fighter and the I.A.R.81 was a multi-role dive bomber who operated as a fighter after dropping its bombs. The aircraft was constantly revised and improved in the factory, which in turn led to a bewildering range of types and versions, each with its on quirks and features. The engineers and designers concentrated mostly on the armament which became increasingly more powerful, but the engine stayed the same due to shortages of supplies and eventually this was to be the 'Achilles' heel' of this aircraft and led to its obsolescence as the war went on. One can only imagine what this aircraft would have looked like and performed with a more powerful engine.

One will see a large number of crashed aircraft photos in this book. That was due to the fact that the factory and the military authorities insisted that every aircraft loss be thoroughly investigated and photographed in as much as possible. Although they are not the most flattering, these photographs are invaluable to the technical researcher because of the wealth of detailed technical information that they provide. This book is leaning more towards the technical aspects of the aircraft rather than a piece of historical research, but I hope that it answers some questions for a wide spectrum of aviation enthusiasts ranging from the historian to the most demanding modeller. Rivet counters may be pleased to know that the price list specifies that as of 1941, a total of 7.315 rivets were used on the I.A.R.80.

Radu Nicolae Brînzan, October 2011

I.A.R. Brașov – History

Eng. Mircea Grosu-Viziru was the driving force behind the I.A.R.80 project, a feat for which he was awarded Ordinul Virtutea Militară cl.III–a (Order of Military Virtue, Third Class).
Photo courtesy of Dan Antoniu

I.A.R. Brașov was inaugurated on 11 October 1927 as a joint venture between the Romanian Government and the French companies Lorraine Dietrich and Bleriot Spad. At first, the factory manufactured aircraft and engines under licence, but in the following years, I.A.R. began to develop their own designs created by an in-house design department.

The first aircraft manufactured by I.A.R. Brașov was the licence-built I.A.R. Morane Saulnier 35 in 1927 and this was followed by the licence-built Potez XXV from 1929. In 1930, the I.A.R-11CV was the first aircraft developed by the in-house design department led by Elie Carafoli, and this aircraft was further developed into a number of subsequent types, but none were put into significant series production. The factory continued to develop aircraft, but as they failed to deliver commercially, they turned towards further licence-built aircraft such as I.A.R. F-10G, (licence-built Consolidated Fleet F-10G) in 1936, the I.A.R. P-11F (licence-built PZL P-11F) in 1937, the I.A.R. P-24E (licence-built PZL P-24) in 1937, I.A.R. S-62bis (licence-built Savoia S-62) in 1938.

In 1938, the company was nationalised and became Regia Autonomă IAR [IAR Independent Board]. The Government used a special air fund to raise capital and soon I.A.R. Brașov turned a corner and began the production of a number of commercially successful aircraft such as the I.A.R.-27, a low-wing monoplane trainer derived from the I.A.R. F-10G and the I.A.R.-37 a light reconnaissance three-seat biplane inspired by the Potez XXV, which was subsequently developed into the I.A.R.38 and I.A.R.39. The factory also continued to manufacture licence-built aircraft such as the I.A.R. FN-305 (licence-built Nardi FN-305) in 1939 and the I.A.R. SM-79 (licence-built Savoia Marchetti SM-79 twin-engine bomber) that was subsequently produced in a number of variants with various engines.

The most successful aircraft, and the one produced in the largest numbers by I.A.R. was the I.A.R.80/81 which began production in 1940.

The I.A.R.47 was to be the last in-house design developed by Regia Autonomă I.A.R. Brașov and this was a low-wing monoplane inspired by the I.A.R.-39. A prototype was built in 1941, but it never reached production. In 1942, I.A.R. Brașov began production of the licence-built Messerschmitt Bf.109 Ga-4 and Bf.109 Ga-6.

A number of American bombing raids in the early half of 1944 culminating with the attacks on 16 April and 6 May caused

Aerial picture of I.A.R. Brașov factory in 1927. The diamond-shaped field in the background is the factory airfield used for reception flights. The large building on the left of the picture is the airframe factory and the hangar bearing the large I.A.R. Brașov name in white was the final assembly hall.
Photo courtesy of Dan Antoniu

In 1930, I.A.R.11 was the first aircraft designed and developed in-house by I.A.R. Braşov. The only prototype built had a wooden structure with aluminium-clad forward fuselage and fabric-covered rear fuselage and wings. The aircraft was powered by a 600HP Lorraine 12FA Courlis water-cooled engine and subsequently served as a basis for the development of a number of different airframes.
Photo courtesy of Dan Antoniu

extensive damage to I.A.R. Braşov and brought an end to large-scale aviation production there. Subsequently, production of Bf.109 was transferred to Caransebeş where it ceased in 1948. When World War II ended, I.A.R. Braşov was converted into a tractor factory called Sovrom Tractor and the next product to leave the production line in 1946 was the I.A.R.-22 tractor.

I.A.R. P-11F was the licence-built version of the Polish P.Z.L. P-11 powered by the 600HP IAR-9K engine. Ninety-five aircraft were manufactured between 1937 and 1938. Some of these aircraft saw action on the Eastern Front after which they were relegated to training duties. This aircraft was fitted with snow skis designed in Romania. The device slung under the cockpit appears to be a gun camera.
Photo courtesy of Dan Antoniu

However, two aircraft workshops were still allowed to function within the factory under the name of ARMV-3 and a design bureau based there produced a number of aircraft in the following years, all of them designed by Radu Manicatide I.A.R.-811 in 1949, I.A.R.-813 in 1950, I.A.R.-814 in 1953 and the MR-2 in 1956. In November 1959, all aircraft production at this I.A.R. Braşov site ceased and a new I.A.R. company was opened in Ghimbav, near Braşov, and still operates to this day.

I.A.R. P-24E was the licence-built version of the Polish P.Z.L. P-24 powered by the 870HP IAR-14K II C32 engine. Twenty-five aircraft were manufactured in 1937 and the experience gained in the manufacture of the I.A.R. P-11 and P-24 was very valuable to the I.A.R. Braşov design bureau, which led to the creation of the I.A.R.80.
Photo courtesy of Dan Antoniu

The new fighter was called 'I.A.R.80' because it was the next aircraft manufactured after the I.A.R. SM-79. The Savoia-Marchetti SM 79 licence was purchased by I.A.R. at the time when the new fighter was proposed.
Photo courtesy of Horia Stoica

Prototype

I.A.R.80 Proposed Specification

At the end of November 1936, I.A.R. proposed to M.A.M. a project for a new all-metal monoplane fighter called 'I.A.R.80'. The aircraft was designed by professors Ion Grosu and Ion Colereanu and engineers Gheorghe Zotta, Mircea Grossu-Viziru and Ioan Wallner. The proposed I.A.R.80 was a single-engine single-seat low-wing monoplane fighter powered by one 930hp in-line inverted-V liquid-cooled Junkers Jumo 211Da engine driving a 3-metre diameter three-blade propeller with electrically adjustable pitch. The cantilever wing was trapeze-shaped with rounded wingtips, 10-metre span, NACA–230412 profile, 17,5% thickness at the wing root decreasing to 5,5% at the tips, 4.10 degree dihedral along the lower surface, 2 degree angle of attack. The all-metal wing was built around two Dural spars and covered with Dural skin. The fabric-covered Handley-Page ailerons had a metal structure and a range of movement between 26 degrees up and 24 degrees down, driven by pushrods linked directly via bell-cranks to the control stick. The fabric-covered Handley-Page flaps had a metal structure and were hydraulically powered, extendable to 45 degrees.

The fuselage had an oval cross-section built in two distinct sections. The front section was built around a welded Chrome Molybdenum tube framework and covered with Dural sheet, housing one upper 263-litre fuel tank, one lower 292-litre fuel tank, one 48-litre oil tank, electric, pneumatic, fuel and oil systems and their fittings and controls. The rear semi-monocoque section was built on stringers, ribs and half-ribs, covered with Dural sheet housing an open cockpit located behind the wing at the front of the semi-monocoque section.

The tail plane and the tail fin were built with metal structures, covered with Dural sheet. The fabric-covered elevators were built with a metal structure and had a range of movement between 25 degrees up and 30 degrees down. The all-metal rudder had a range of movement of up to 25 degrees left and right. The retractable undercarriage was a Messier design with oleo-pneumatic dampeners; it had a 3.45 metre wheel track and was fitted with 635 x 190 mm wheels and hydraulic brakes. Armament consisted of 2 x 7,92mm FN Browning machine located in the wings outside the propeller arc.

Specifications

- Wingspan — 10,000 mm / 32.80 ft
- Length — 8,850 mm / 29.03 ft
- Lift area — 15.50 m2 / 166.84 sq.ft.
- Tail plane area — 2.730 m2 / 29.38 sq.ft.
- Fin area — 1.161 m2 / 12.49 sq.ft.
- Aileron area — 2x0.680 m2 / 7.32 sq.ft.
- Flap area — 2x0.680 m2 / 7.32 sq.ft.
- Empty weight (incl. three-blade propeller and radio) — 1845 kg. / 4067.52 lb
- Max. gross weight — 2470 kg. / 5445.41 lb
- Max. speed at 5700 m — 558 km/h. / 346.72 mph

Prototype

As proposed in the specification, the prototype was to be powered by one 930 hp Junkers JuMo 211Da engine. The tail from frame 5 to the rudder was based on modified and strengthened parts taken from a P.Z.L. P-24 manufactured by I.A.R. Braşov under licence at the time. The rest of the fuselage including the cockpit was a new design as well as the modern low single wing and retractable landing gear. Construction commenced in 1937 and was completed by the end of the year. Even though MAM ordered three JuMo 211Da engines from Germany in 1937, two intended for the Savoia SM-79B prototype and one for the I.A.R.80 prototype, for various reasons the delivery of these engines continued to be delayed. This caused a significant setback for the completion of the prototype and in early-1936, I.A.R. was asked to redesign the prototype to be powered by one 870 hp I.A.R.-14K IIc32 engine, the most powerful engine available in Romania at the time. This engine was 200 kg lighter than the JuMo 211Da engine, which compensated for the lower power output but it consumed a lot of oil and had a time-between-overhaul of only 30 hours. The prototype equipped with the new engine was completed in the spring of 1939. The design and construction work totalled 180,000 man-hours. On 4 April 1939, the prototype flown by Cpt.Av. Dumitru Popescu took-off in a short factory test flight. The official inaugural flight was carried out by Cpt.Av. Dumitru Popescu on 12 April 1939. In early-May 1939, the aircraft was flown from Braşov to Escadrila de Experienţe Pipera where the serial number '0' was reportedly painted on the tail fin. MAM insisted that the reception flight be carried out by an experienced pilot and for this purpose they chose the French pilot Michel Detroyat who was an accomplished test and aerobatic pilot. On 17 May 1939, representatives of Ministerul Aerului şi Marinei, I.A.R. Braşov factory, staff of Escadrila de Experienţe and some of the staff of Pipera airfield gathered to watch the official reception flight carried out by Michel Detroyat. After the flight, while taxiing, the aircraft became bogged in a spot of soft ground and nosed over breaking the propeller. After this incident, the aircraft was dismantled and sent to I.A.R. Braşov by train where it was reassembled, fitted with a wooden propeller and continued the test programme. During the subsequent tests, the pilots criticised the low engine output and the poor visibility from the cockpit while taxiing or landing, but appreciated the flight characteristics of the aircraft. On 20 July 1939, Feast Day of Saint Elijah, the Patron Saint of Romanian Aviators, the I.A.R.80 prototype was presented to the public for the first time at an aviation meeting held on Cotroceni airfield when the silver-coloured plane decorated with a stepped red arrow carried out a high speed pass in front of the official tribune.

©Teodor Liviu Moroşanu 2010

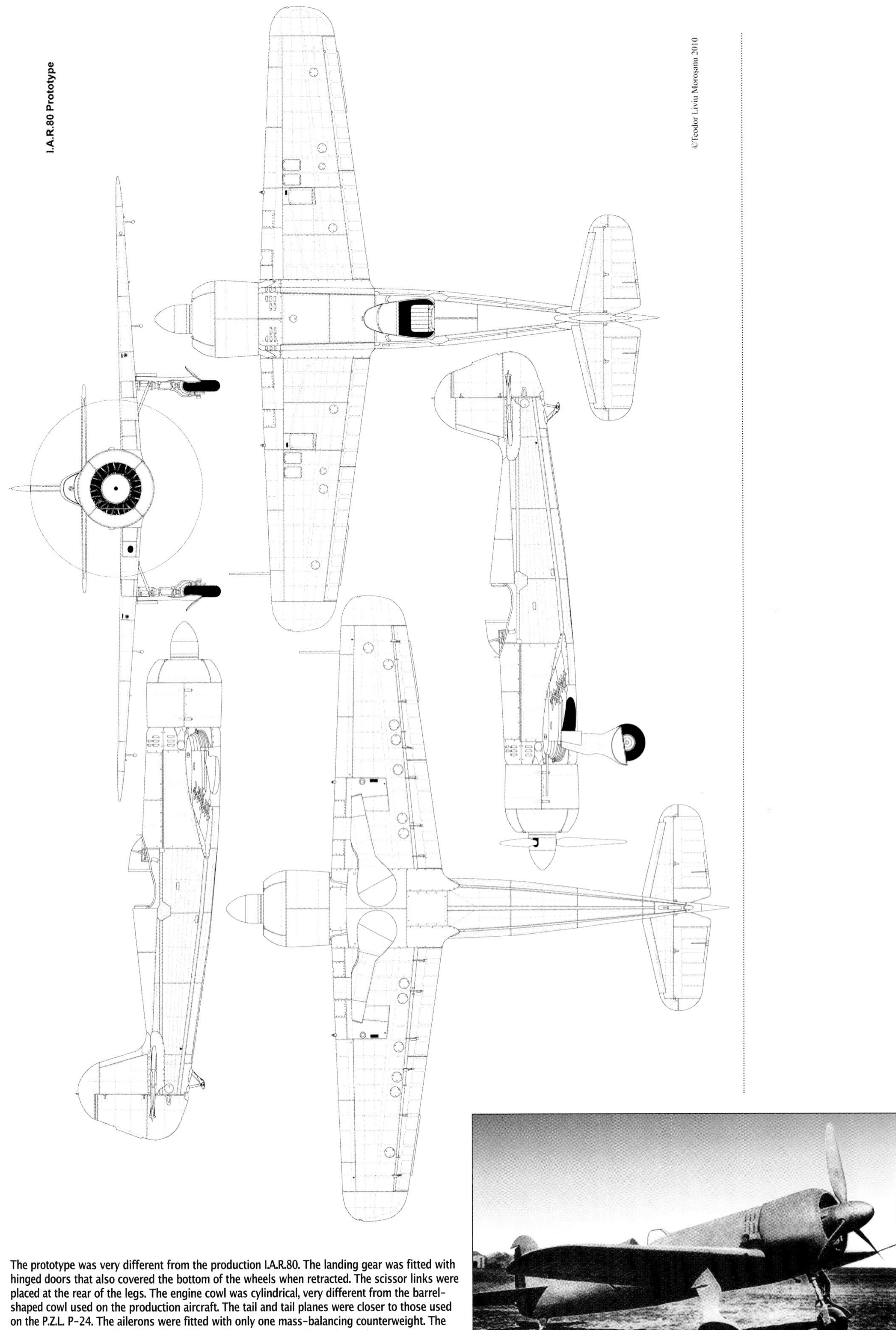

The prototype was very different from the production I.A.R.80. The landing gear was fitted with hinged doors that also covered the bottom of the wheels when retracted. The scissor links were placed at the rear of the legs. The engine cowl was cylindrical, very different from the barrel-shaped cowl used on the production aircraft. The tail and tail planes were closer to those used on the P.Z.L. P-24. The ailerons were fitted with only one mass-balancing counterweight. The oil cooler intake in the leading edge of the right wing was smaller and round.

Photo courtesy of ABC Collection

I.A.R.80 No. 1 to 20

I.A.R.80 No. 1 to 20

By the end of 1939, the entire test programme of the prototype was completed, including aerial combat trials. The modifications proposed by the test pilots were analysed by MAM, Serviciul Tehnic of SSA and I.A.R. Braşov, which led to the changes implemented in the series aircraft. As a result, MAM placed Contract No.2027/1939 for the construction of fifty I.A.R.80 fighters. No.1 was the first production aircraft and was significantly different from the prototype. The engine cowl was significantly modified and was fitted with variable cooling gills that controlled the amount of cooling air passing over the cylinders. The cross-section of the cowl was changed, round at the front and oval at the rear. Three new exhaust pipes were fitted, two fitted to front-row cylinders No. 4 and No. 5, venting underside on the left side of the supercharger intake and one fitted to back-row cylinder No.4 venting underside on the right side of the supercharger intake. The engine mount was extended and as a result the fuselage was extended by 74 mm. The supercharger was fitted with an 'elbow' intake protruding below the engine cowl. Slots were fitted to the top of the forward fuselage ahead of the firewall, to aid engine cooling and venting some of the heat from the oil radiator and oil tank. The cockpit was moved aft and fitted with an enclosed Plexiglas canopy and a new windscreen.

The rear fuselage spine was completely redesigned and enlarged. The structure of the tail planes was strengthened and the support struts were removed. The elevators were fitted with adjustable trim tabs controlled from the cockpit. The rudder was redesigned and was fitted with a fixed trim tab. A white navigation light was fitted to the top of the fin. The wingspan was extended by 520 mm. The ailerons were extended and were fitted with another counterweight and pivot point. Navigation lights were fitted to the wingtips. The intake in the root of the right wing leading edge for the oil radiator was enlarged to improve oil cooling. The armament was improved by the addition of two further 7.92 mm FN Browning machine guns, increasing the firepower to four machine guns. The ammunition was increased to 2,400 rounds and the access panels for the ammunition and

I.A.R.80 No. 1 and I.A.R.80-B No. 227 photographed while flying over the Carpathian Mountains in the autumn of 1942. I.A.R.80 No.1 was used for extensive trials by Escadrila de Experienţe between September and December 1940. These tests revealed a number of shortcomings that were corrected in subsequent airframes. This aircraft was lost while flown by Adj.Av. Ion Ionescu on 18 June 1943 when the horizontal tail-plane broke-off in flight and the plane crashed near the locality of I.C. Brătianu. At that time, this aircraft was part of Şcoala de Vânătoare, Flotila 3 Vânătoare Galaţi.
Photo courtesy of ABC Collection

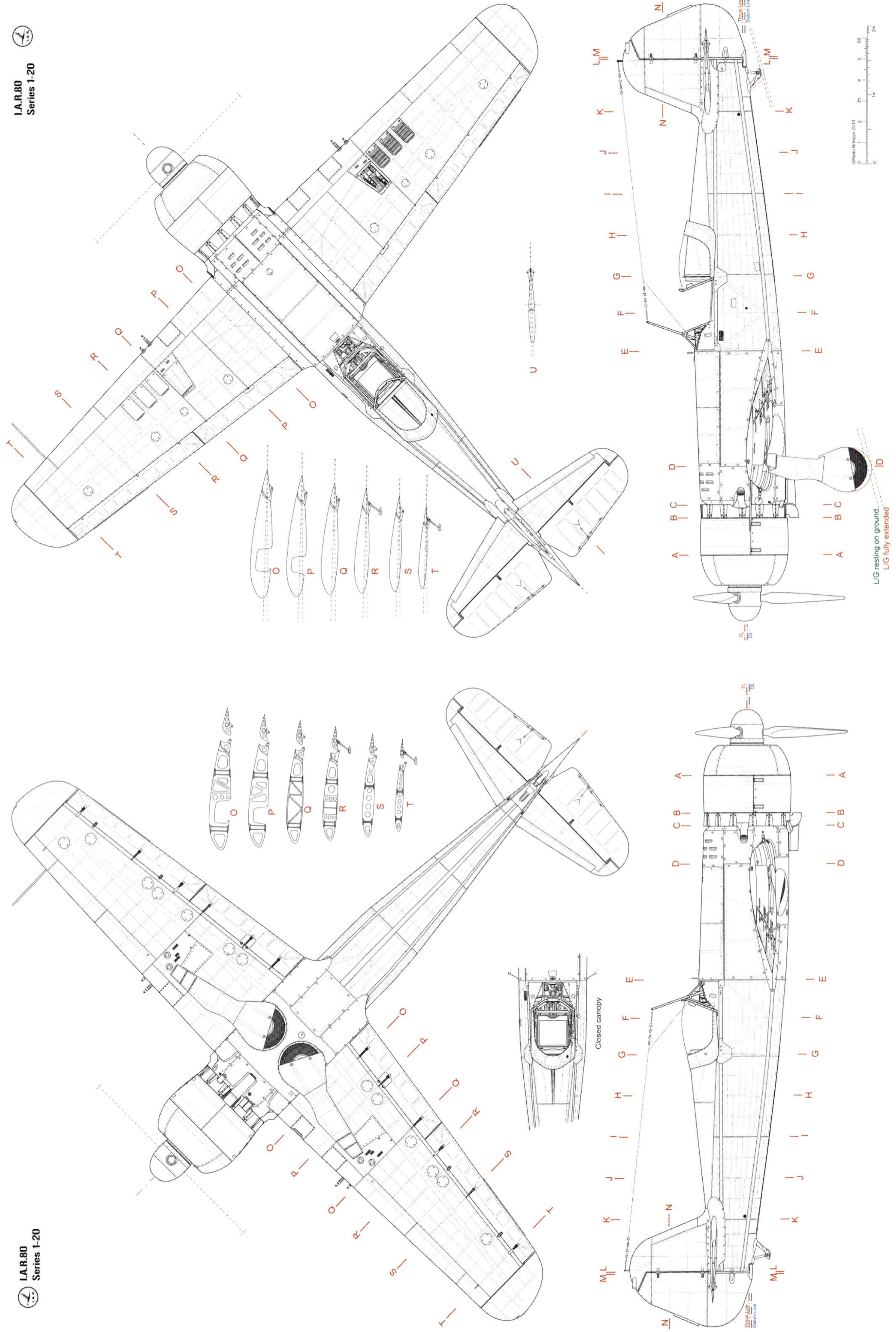
U
Closed canopy
L/G resting on ground
L/G fully extended
©Radu Brinzan 2010

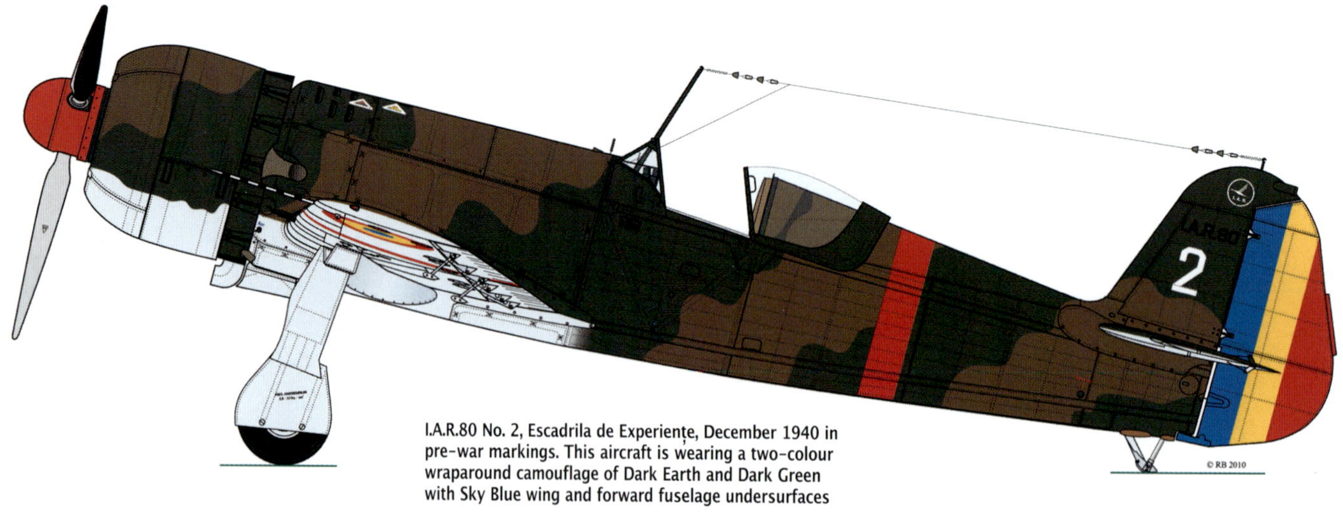

I.A.R.80 No. 2, Escadrila de Experiențe, December 1940 in pre-war markings. This aircraft is wearing a two-colour wraparound camouflage of Dark Earth and Dark Green with Sky Blue wing and forward fuselage undersurfaces

I.A.R.80 No. 3, Școala de Vânătoare, Flotila 3 Vânătoare, Galați, October 1943. This aircraft is wearing a single-colour uppersurface camouflage of Olive Green (possibly RLM64) with Light Blue (possibly RLM76) undersurfaces. Eastern Front markings

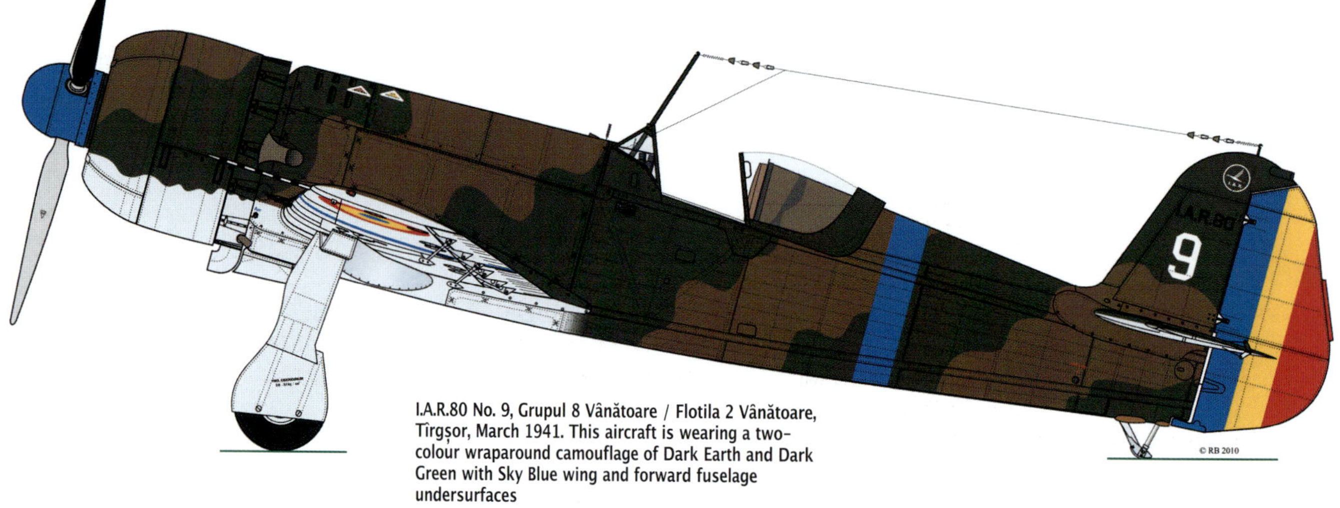

I.A.R.80 No. 9, Grupul 8 Vânătoare / Flotila 2 Vânătoare, Tîrgșor, March 1941. This aircraft is wearing a two-colour wraparound camouflage of Dark Earth and Dark Green with Sky Blue wing and forward fuselage undersurfaces

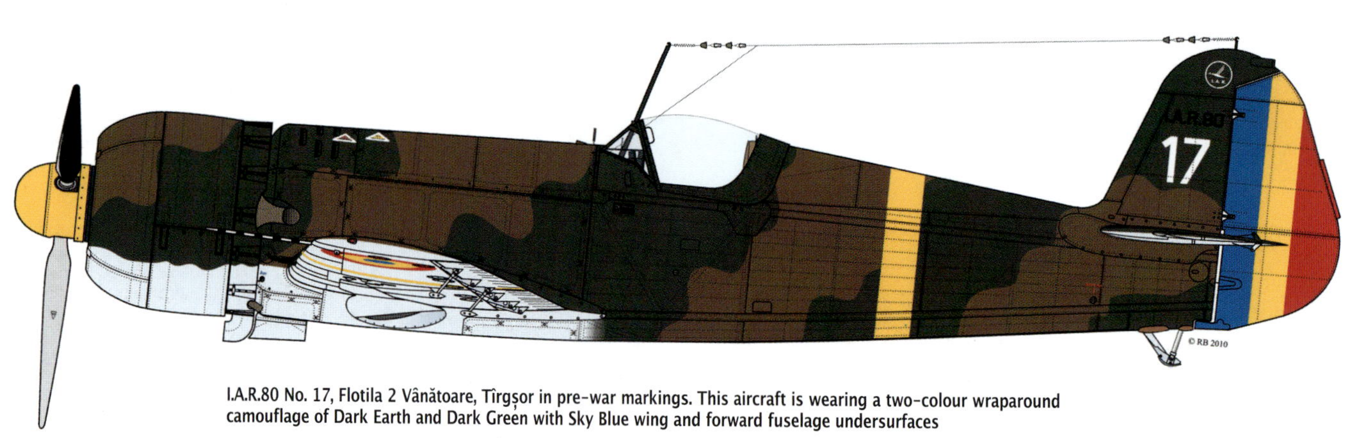

I.A.R.80 No. 17, Flotila 2 Vânătoare, Tîrgșor in pre-war markings. This aircraft is wearing a two-colour wraparound camouflage of Dark Earth and Dark Green with Sky Blue wing and forward fuselage undersurfaces

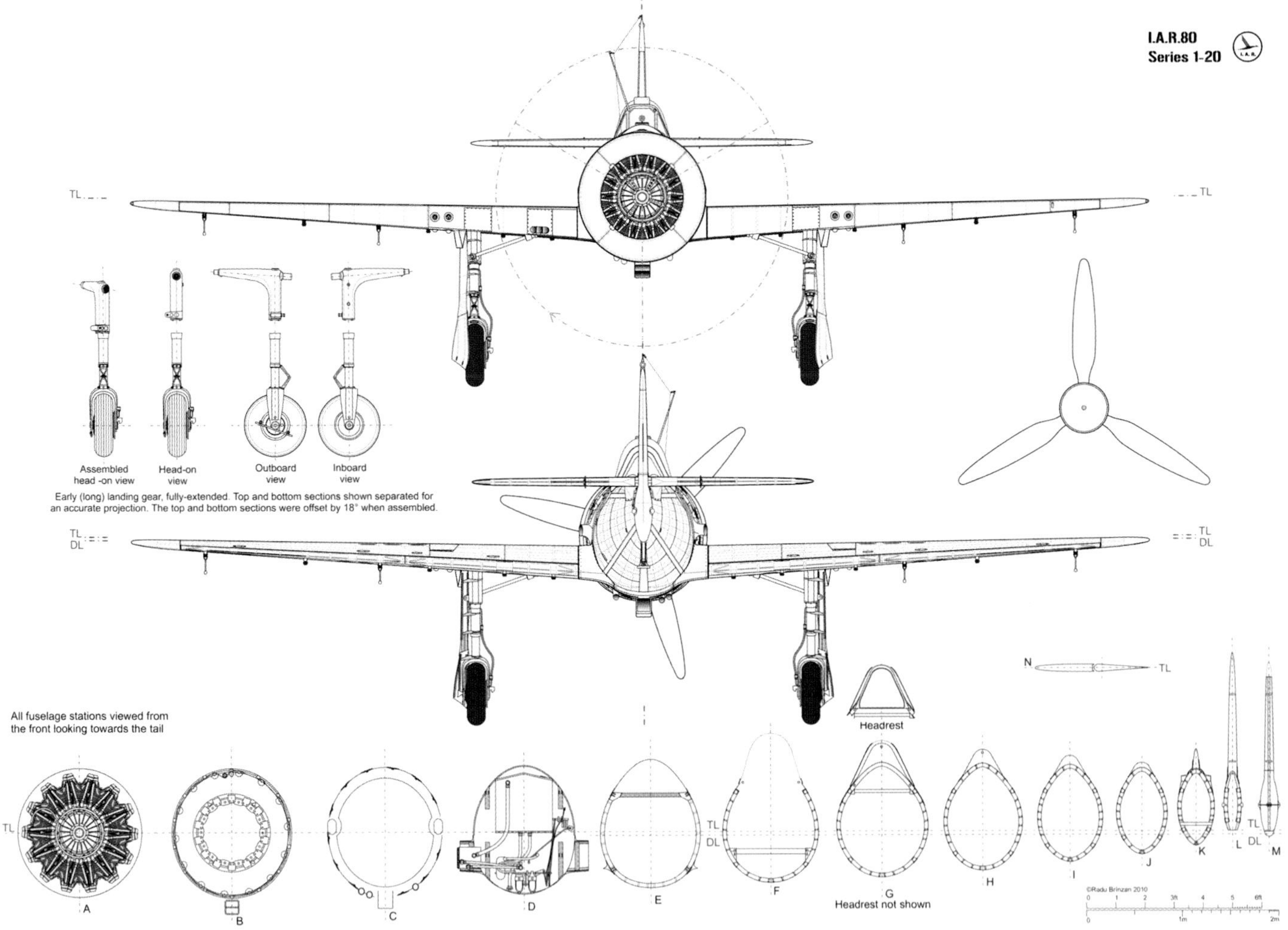

Early (long) landing gear, fully-extended. Top and bottom sections shown separated for
an accurate projection. The top and bottom sections were offset by 18° when assembled.

Assembled head -on view — Head-on view — Outboard view — Inboard view

All fuselage stations viewed from
the front looking towards the tail

Headrest

Headrest not shown

©Radu Brinzan 2010

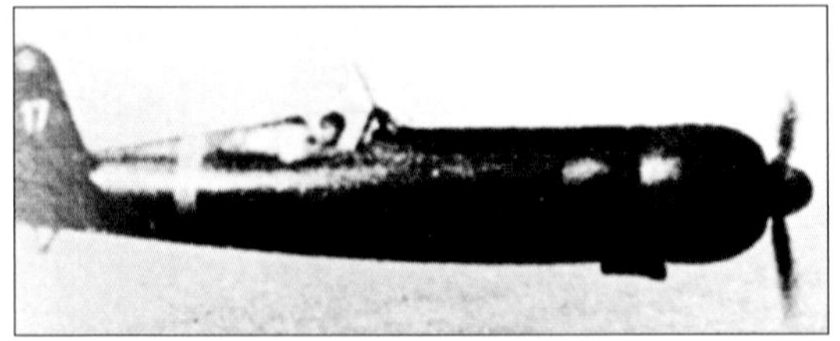

Between 1 March and 11 April 1941, I.A.R.80 No. 17 of Fl.2Vt. was sent
to Mamaia on the Black Sea coast where it was used by a variety of
pilots to practice aerial gunnery. Note the absence of the rear-view
mirror on the canopy and the pre-war markings Photo courtesy of ABC Collection

At 16.15 on 15 October
1943, Elev Av. Ciobănaș
Andrei of Șc.Vt./Flt.3Vt.
Galați belly-landed I.A.R.80
No. 3 near Tighina airfield
due to a seized engine.
The aircraft was repaired
and flew again. The
aircraft appears to be
equipped with only one
machine gun in each wing.
The camouflage paint on
the top of the fuselage
was damaged by fuel
spilled during refuelling.
This aircraft was
retrofitted with the
external reinforcement
fuselage brace introduced
starting with No. 21. Note
the late-type headrest-
mounted seatbelts
introduced in May 1943
and the late single-colour
upper surface camouflage
scheme
Photo courtesy of Pitești Military
Archives via ABC Collection

guns were modified. The guns could be charged only on the
ground and were fired pneumatically. The aircraft was equipped
with a 'ring and bead' gunsight and was fitted with a Telefunken
FuG VIIa radio. The cockpit was fitted with a French
Munerele-34 oxygen regulator. A manually operated fire
extinguisher was also fitted. The landing gear scissor links were
mounted at the front of the landing gear legs. The landing gear

doors were modified by removing the hinged lower flap (thus
leaving the bottom one-third of each wheel exposed) and the
main door was made in two parts, the upper part attached to the
leg and the lower part attached to the wheel fork. As the war
progressed, the armament fitted to the aircraft of this series
became increasingly obsolete and most of them were relegated to
aviation schools where they were used for training.

At 10.50 on 8 March 1941,
Adj.Stg.Av. Vasile Niță of
Gr.8Vt./Flt.2Vt. crash-landed
I.A.R.80 No. 9 while landing
on Tîrgșor airfield. The
aircraft suffered minor
damage and was repaired by
I.A.R. Brașov. Note the
natural metal propeller and
the pre-war markings
Photo courtesy of Pitești Military Archives
via ABC Collection.

I.A.R.80 No. 21 to 50

One of the most famous photographs of an I.A.R.80 is the tail of No. 42 decorated with a tally of thirteen victories. This photograph was published in Romanian, German and Italian propaganda magazines. The tally does not represent the individual score of any pilot, but rather it represents the victories claimed by the entire unit. This photograph was reportedly taken on the Eastern Front during the summer of 1941. According to the aircraft records, at that time, I.A.R. No. 42 was part of Esc.42/52Vt. based in Focșani in July, in Bîrlad and Sturzeni in August and in Salz in September. In September 1941, this aircraft was sent back to I.A.R. Brașov for overhaul and in October it joined Esc.60Vt./Gr.8Vt. in Mizil
Photo courtesy of ABC Collection

Following service use, a number of changes were implemented starting with No. 21. In order to limit airframe stress, an external brace was riveted on airframe 5, just behind the cockpit. This brace was subsequently fitted to all airframes until No. 94. Starting with No. 21, the 'ring and bead' gunsight was replaced with a Goerz reflex gunsight. All of these changes were retrofitted to the previous airframes. As the war progressed, the armament fitted to the aircraft of this series became increasingly obsolete and most of them were relegated to aviation schools where they were used for training.

In July 1941, I.A.R.80 No. 22 of Flt.2 Vt./Gr.1 Vt., Esc.42/52 Vt. was flown by Adj.Stag.Av. Gheorghe Firimide from Bîrlad Airfield escorting bombers during the early days of the Bessarabian Campaign. A specific feature of the theatre markings applied to the noses of aircraft during the early days of the Bessarabian Front was a yellow segment applied only to the underside of the green engine cowl. The Italian Macchi MC200 in the foreground was flown by Maurizio Ruspoli of Poggio Suasa, a Regia Aeronautica 10-victory ace who flew on the Eastern Front with Corpo di Spedizione Italiano in Russia.
Photo courtesy of ABC Collection

On 18 September 1941, Adj.Stag.Av. Ion Oprea capsized I.A.R.80 No. 22 of Escadrila 42/52 Vânătoare, after hitting a pothole while landing on Salz airfield. Note the unpainted frame behind the head armour. This aircraft is wearing the early type of cross
Photo courtesy of ABC Collection

"

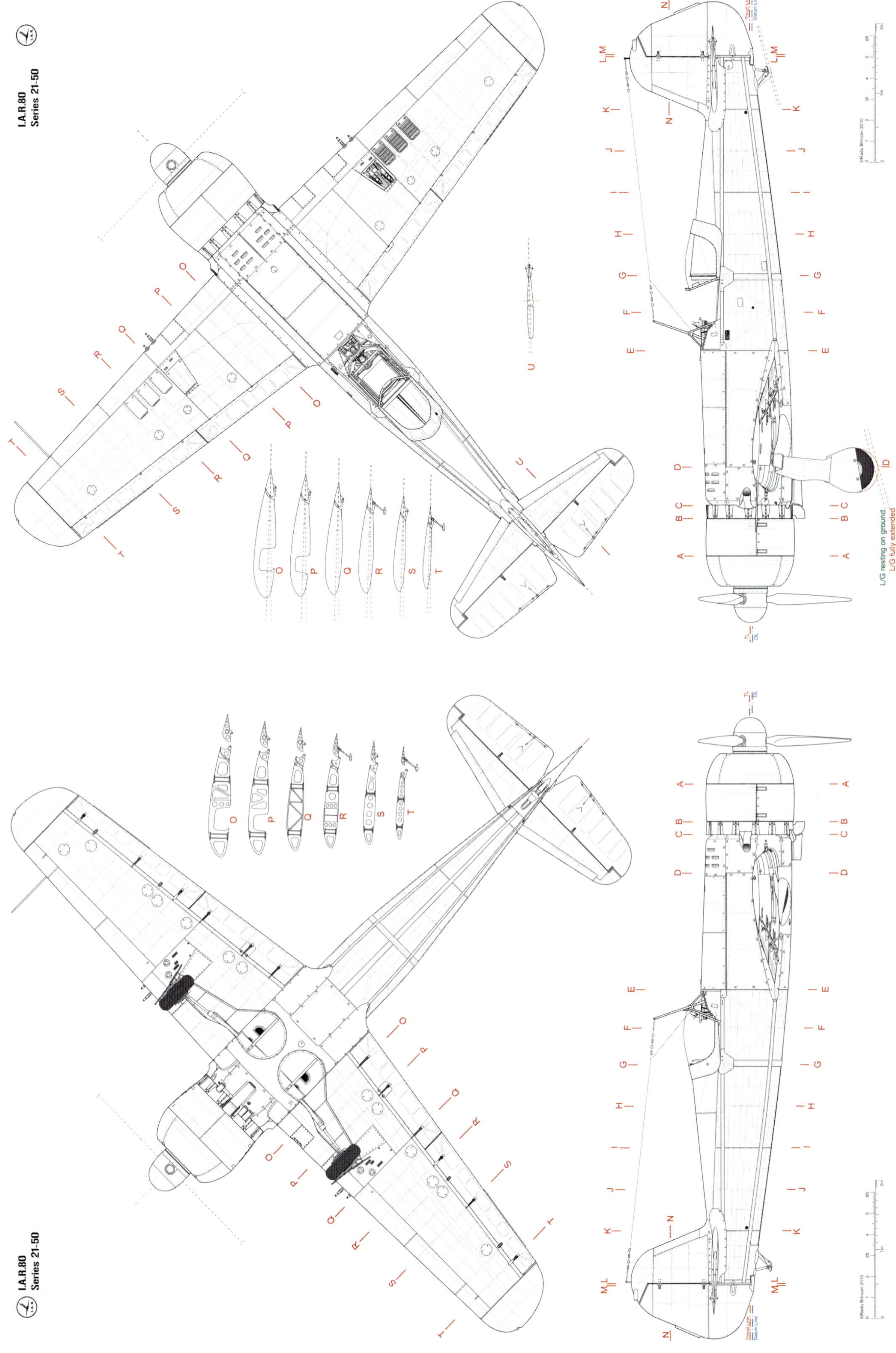

Datum line
Datum Line
N
L M
K
J
I
H
G
F
E
D
C
B
A
N
L M
K
J
I
H
G
F
E
D
C
B
A
L/G resting on ground.
L/G fully extended
©Radu Brinzan 2010
A
C B
D
E
F
G
H
I
J
K
M L
N
A
C B
D
E
F
G
H
I
J
K
M L
N
Datum Line
©Radu Brinzan 2010
P
Q
R
S
T
O
P
Q
R
S
T
O
U
U
O
P
Q
R
S
T
O
P
Q
R
S
T
O
Q
R
S
T

I.A.R.80 No. 22, Flotila 2 Vânătoare, Grupul 1 Vânătoare, Escadrila 42/52 Vânătoare, July 1941, Bîrlad. This aircraft is wearing a single-colour uppersurface camouflage of Dark Green with Sky Blue undersurfaces. Eastern Front markings. The top of the engine cowl was painted Dark Green

I.A.R.80 No. 22 of Escadrila 42/52 Vânătoare, September 1941, Salz. This aircraft is wearing a single-colour wraparound camouflage of Dark Green with Sky Blue undersurfaces. Eastern Front markings. The engine cowl is entirely painted yellow

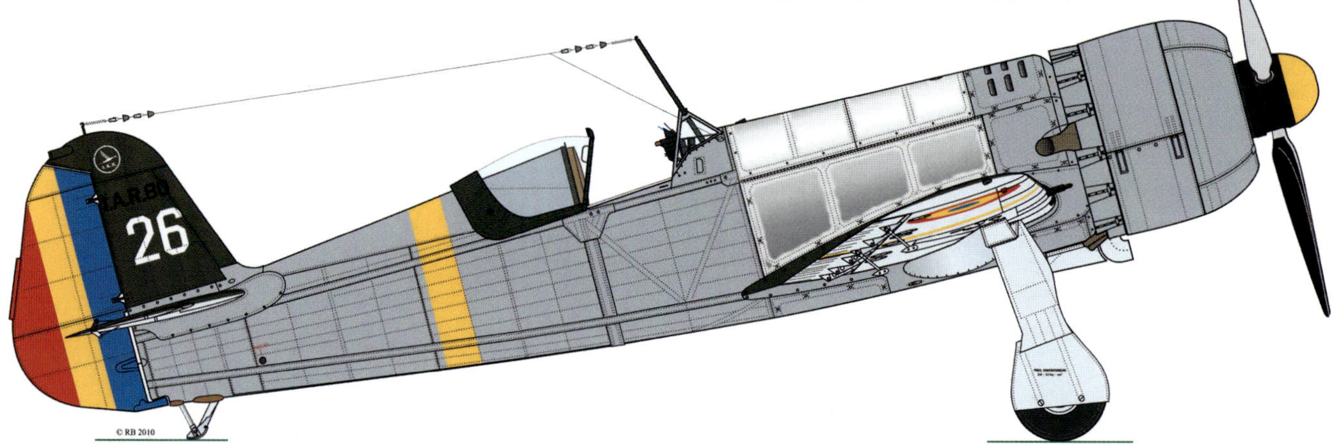

I.A.R.80 No. 26, I.A.R. Brașov, February 1941, in pre-war markings. This aircraft is partially painted. The wings have a two-colour uppersurface camouflage of Dark Earth and Dark Green with Sky Blue undersurfaces. The rear fuselage is painted Cerrux Grey primer. The nose section is unpainted natural metal. The tailfin is painted Dark Green

I.A.R.80 No. 42, Escadrila 42/52 Vânătoare, Grupul 5 Vânătoare Salz airfield, August 1941. This aircraft is wearing a two-colour wraparound camouflage of Dark Earth and Dark Green with Sky Blue wing and forward fuselage undersurfaces. Eastern Front markings

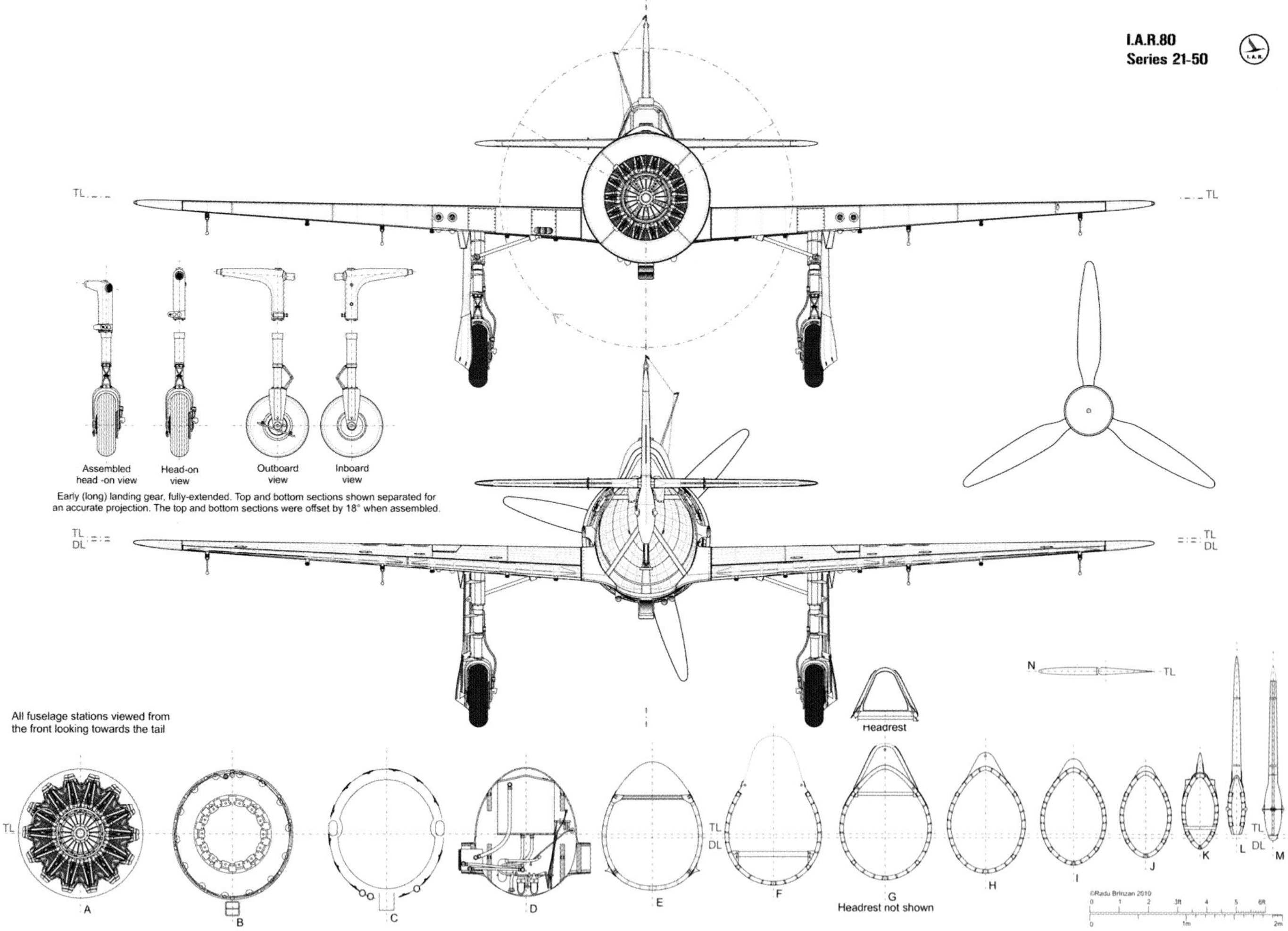

In March 1941, I.A.R.80 No. 26 was sent to Escadrila de Experiențe on Pipera airfield where it was used alongside No. 1 and No. 2 for trials and tests aimed at testing the improvements and modifications suggested by pilots. In this picture, the aircraft is still in the factory, alongside other aircraft awaiting completion. The aircraft is unpainted with large areas of the fuselage in primer or natural metal, but it is already wearing pre-war markings. On 13 July 1941, when it was part of Esc.60Vt./Gr.8Vt. this aircraft was lost when it collided with I.A.R.80-A No. 52 while scrambling from Bârlad airfield with Adj.Stg.Av. Ioan Stanciu at the controls.
Photo courtesy of ABC Collection

I.A.R.80 No. 26 of Esc.60Vt./Gr.8Vt. photographed on Sturzeni airfield in August 1941. This photograph was taken with orthochromatic film and as a result, the yellow engine cowl and the crosses appear very dark. At 07.45 on 2 October 1941, while taking off from Salz airfield, this aircraft flown by Adj.Av. Gheorghe Țifrea collided with I.A.R.80 No. 16 and both aircraft were destroyed
Photo courtesy of ABC Collection

I.A.R.80-A No. 51 to 75

Starting with No. 51, the armament was improved by the addition of two further 7.92 mm FN Browning machine guns bringing the total firepower to six machine guns. However, the amount of ammunition remained the same as in the previous series, respectively 2,400 rounds, 400 rounds per gun. A new, taller armoured backrest was fitted to the seat. Because of difficulties with the purchase and delivery of Goerz gunsights from Austria, a Telereflex-type gunsight manufactured under licence in Romania by IOR was introduced beginning with this series, but existing stocks of Goerz gunsights continued to be used. A wooden propeller was tested on No.58 but the results were not satisfactory. This aircraft was also tested with a longer control stick, but the results were also unsatisfactory. As the war progressed, the armament fitted to the aircraft of this series became increasingly obsolete and most of them were relegated to aviation schools where they were used for training.

On 21 August 1941, I.A.R.80-A No. 75 of Esc.59Vt./Gr.8Vt. with Adj.Av. Vasile Mirilă at the controls, capsized while landing on Mamaia airfield due to the fracture of the landing gear scissor links. The aircraft suffered little damage, was repaired and flew again. Note the green camouflage applied to the top of the engine cowl *Photo courtesy of ABC Collection*

I.A.R.80-A No.61, Școala de Vânătoare, Escadrila 44 Vânătoare, Flotila 3 Vânătoare, Galați, August 1943. This aircraft is wearing a single-colour uppersurface camouflage of Olive Green (possibly RLM64) with Light Blue (possibly RLM76) undersurfaces. Eastern Front markings

I.A.R.80 No. 67, Scoala de Vânătoare Escadrila 44 Vânătoare, Flotila 3 Vânătoare, Galați, June 1943. This aircraft is wearing a single-colour uppersurface camouflage of Olive Green (possibly RLM64) with Light Blue (possibly RLM76) undersurfaces. Eastern Front markings

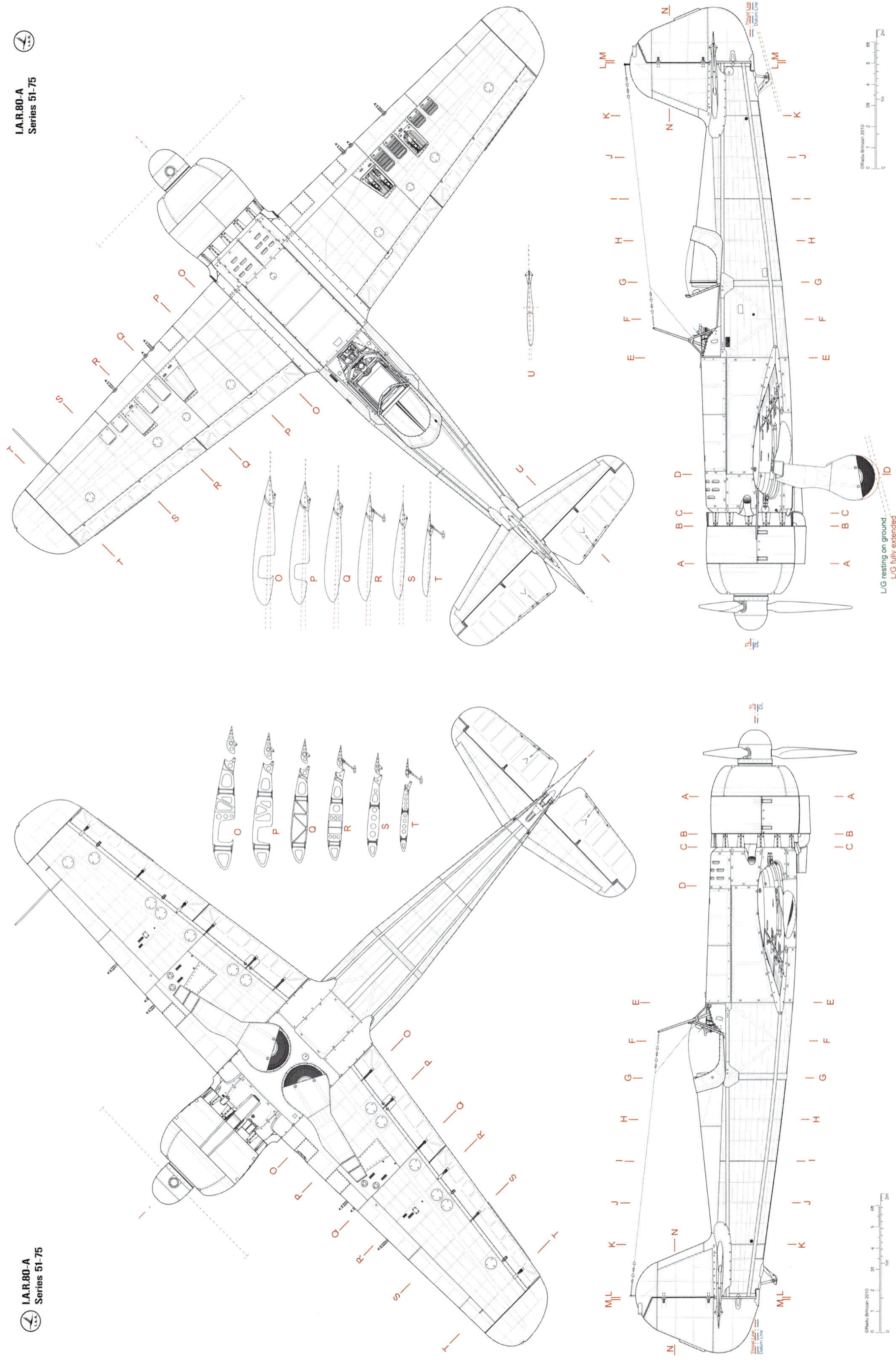
I.A.R.80-A
Series 51-75
I.A.R.80-A
Series 51-75
Thrust Line
Datum Line
L/G resting on ground
L/G fully extended
©Radu Brinzan 2010

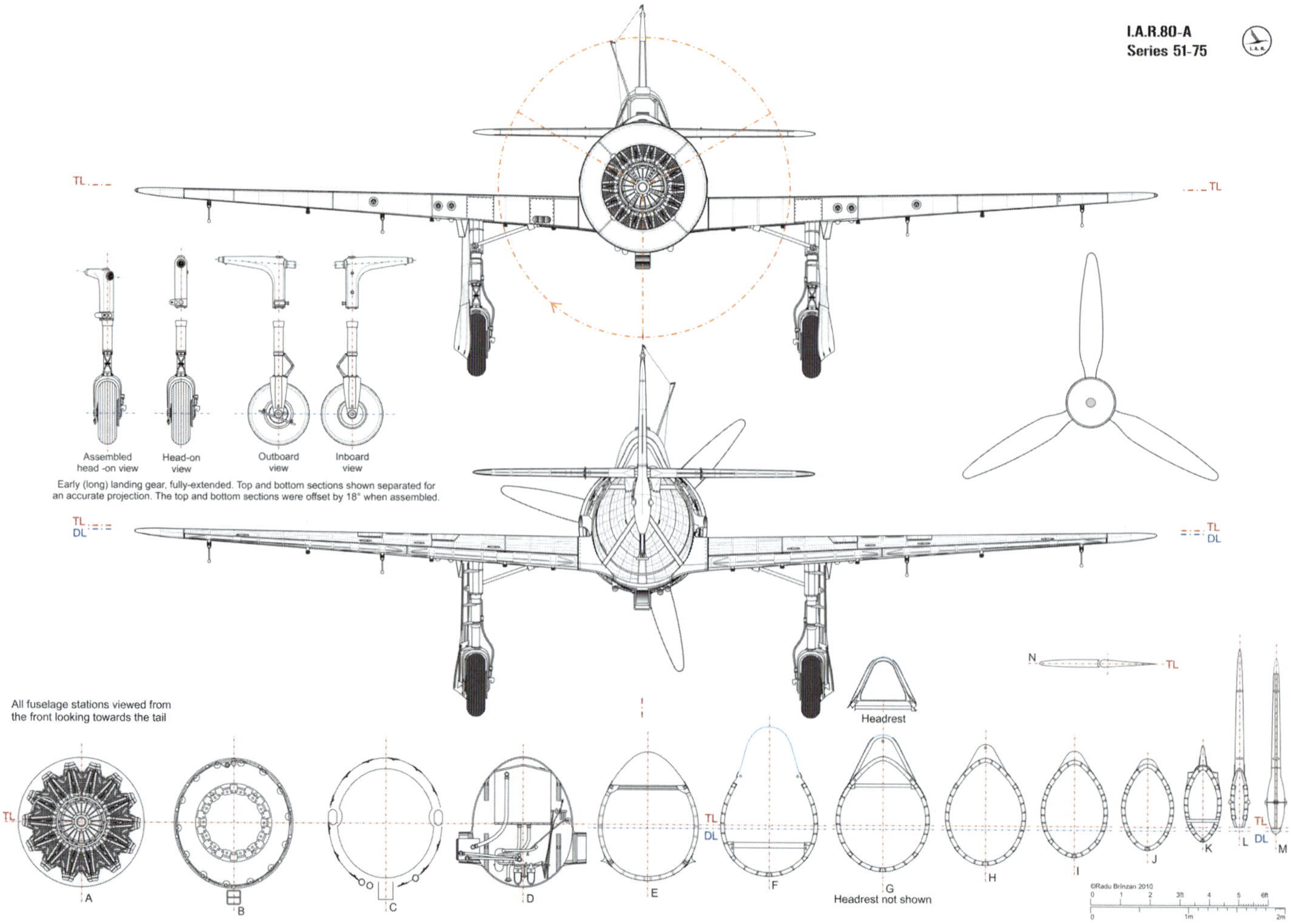

Assembled head-on view
Head-on view
Outboard view
Inboard view

Early (long) landing gear, fully-extended. Top and bottom sections shown separated for an accurate projection. The top and bottom sections were offset by 18° when assembled.

All fuselage stations viewed from the front looking towards the tail

Headrest

Headrest not shown

©Radu Brinzan 2010

A B C D E F G H I J K L M

At 11.30 on 06 August 1943, Elev Av. Valentin Conovici crash-landed I.A.R.80-A No.61 of Școala de Vânătoare Flt.3Vt. Galați near the village of Brănești. This aircraft was fitted with a late-type air filter, headrest-mounted late-type seatbelts but was still equipped with a ring and bead gunsight. The camouflage scheme is a single-colour olive green with light blue fuselage underside. The top of the engine cowl was covered with green paint leaving a yellow section at the bottom. Note the long propeller spinner with a white segment
Photo courtesy of Pitești Military Archives via ABC Collection

I.A.R.80-A No. 60 of Școala de Vânătoare Flt.3Vt. photographed on a snowy Galați airfield in early 1943. Note the towing device used to lift the aircraft by the tail skid and wheel it around the airfield.
Photo courtesy of ABC Collection

On 26 June 1943, while trying to land I.A.R.80-A No. 67 of Școala de Vânătoare, Escadrila 44 Vânătoare, Flotila 3 Vânătoare, Galați, August Petre Scurtu landed at stall speed and when he overcompensated during the recovery, the landing gear collapsed damaging the aircraft. The aircraft was repaired by I.A.R. Brașov and flew again
Photo courtesy of Pitești Military Archives via ABC Collection

I.A.R.80-A No. 76 to 90

The 76 – 90 series was used to prepare and implement on the production line the modifications needed for the forthcoming BoPi version. In order to allow for the future installation of a central bomb rack, the wheel wells were placed further apart from each other thus widening the panel between the wheel wells, which was given a slightly bulged shape. In order to allow the landing gear to extend or retract without hitting the centrally-mounted bomb, each of the landing gear oleos were shortened by 60mm. As a result, the ground clearance of the propeller was reduced from 375mm to 300mm and the pilot's view from the cockpit while on the ground was improved but the trade-off was a slightly bumpier run.

Starting with No. 78, the flap mechanisms were modified in order to allow them to be secured in two positions: 45° for take-off and 60° for use as a dive brake. A strengthened type of flaps was devised for use with the BoPi aircraft and introduced with this series. Also, a new type of flap was tested, strengthened by the addition of a long chrome-molybdenum sleeve that the control lever was welded to. However, due to shortages of supplies, this was not introduced until No. 151 and an interim strengthened flap with strengthened ribs and control lever was implemented instead.

Beginning with No.77, the oxygen regulator was replaced by a German Dräger HLa 732 system, which was used on all subsequent series and retrofitted to the existing aircraft. As the war progressed, the armament fitted to the aircraft of this series became increasingly obsolete and most of them were relegated to aviation schools where they were used for training.

Although this aircraft is not clearly identified by the number on the tail, reportedly the aircraft pictured in this photograph is I.A.R.80-A No. 84 and was a school aircraft. According to the aircraft records, I.A.R.80-A No. 84 was part of Şcoala de Vânătoare Flt.3Vt. Galaţi between November 1942 and March 1944 when it was transferred to Esc.50Vt. Ghimbav. The block-letter graffiti on the fuselage reads 'Apasă de mânerul trenului tare spre înapoi şi apoi înainte, apasă pe pedală cu piciorul stâng' meaning 'Push the chassis handle hard backward and then forward, push the pedal with the left foot'. This appears to be an instruction for retracting the landing gear and counteracting take-off tail-swing.

Photo courtesy of ABC Collection

I.A.R.80-A No. 82 and I.A.R.80-A No. 51 of Esc.59Vt./Gr.8Vt. based in Mamaia flew combat missions during August 1944 in the Tulcea-Sulina area, over the Danube Delta. In mid-September 1944, both aircraft were transferred to Esc.42/52Vt. on the Eastern Front, but three days later No. 82 was sent back to Esc.59.Vt. On 10.October.1941, Cpt.Av. Ernest Vernescu, a veteran of forty-three combat missions, was hit by Soviet AA fire over Odessa harbour and crashed in the village of Dalnik. He was flying a scout mission in the Odessa-Liman Tătarca grid in order to assess the number of vessels present in Odessa harbour and their movements. On the way back, his wingman noticed his absence and radioed for his position. Cpt.Av. Vernescu replied 'I am North of Sukhoy Liman, at 20 km' followed by silence. Ground troops reported that they saw a plane, which was not in flames, crashing into the middle of the village

Photo courtesy of ABC collection

I.A.R.80-A No.79 after the crash in July 1944. Note the late-type cooling gills on the fuselage sides behind the exhausts, which may indicate that this aircraft was also retrofitted with a second oil radiator in the left wing

Photo courtesy of Piteşti Military Archives via ABC Collection

At 16.15 on 19 July 1944, I.A.R.80-A No. 79 flown by pilot Mihai Căciulă hit the control station while landing on Ghimbav airfield and was damaged. Note the olive green upper surface camouflage with light blue fuselage underside

Photo courtesy of Piteşti Military Archives via ABC Collection

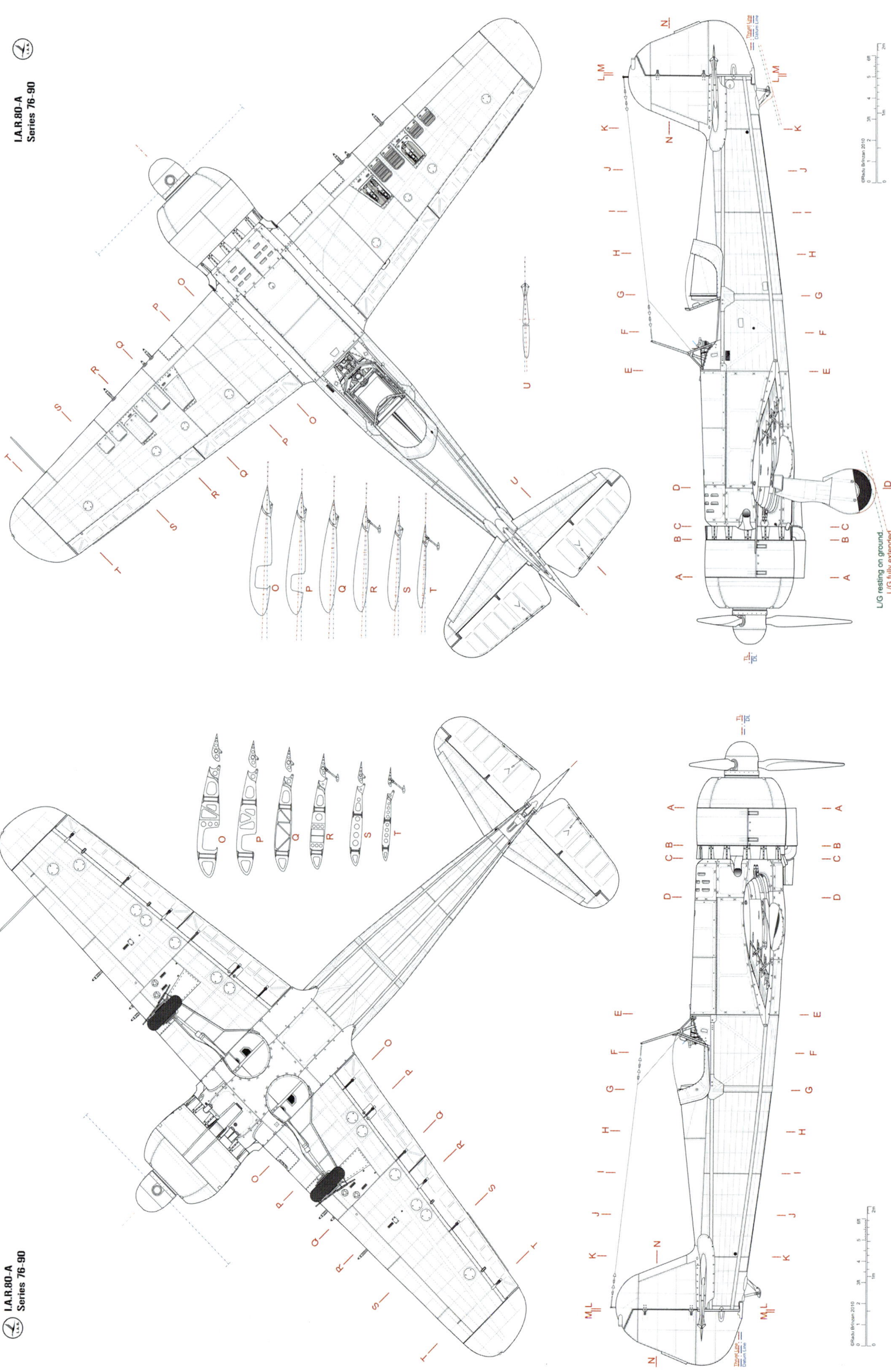
I.A.R.80-A
Series 76-90
Thrust line
Datum Line
L/G resting on ground.
L/G fully extended.
©Radu Brînzan 2010

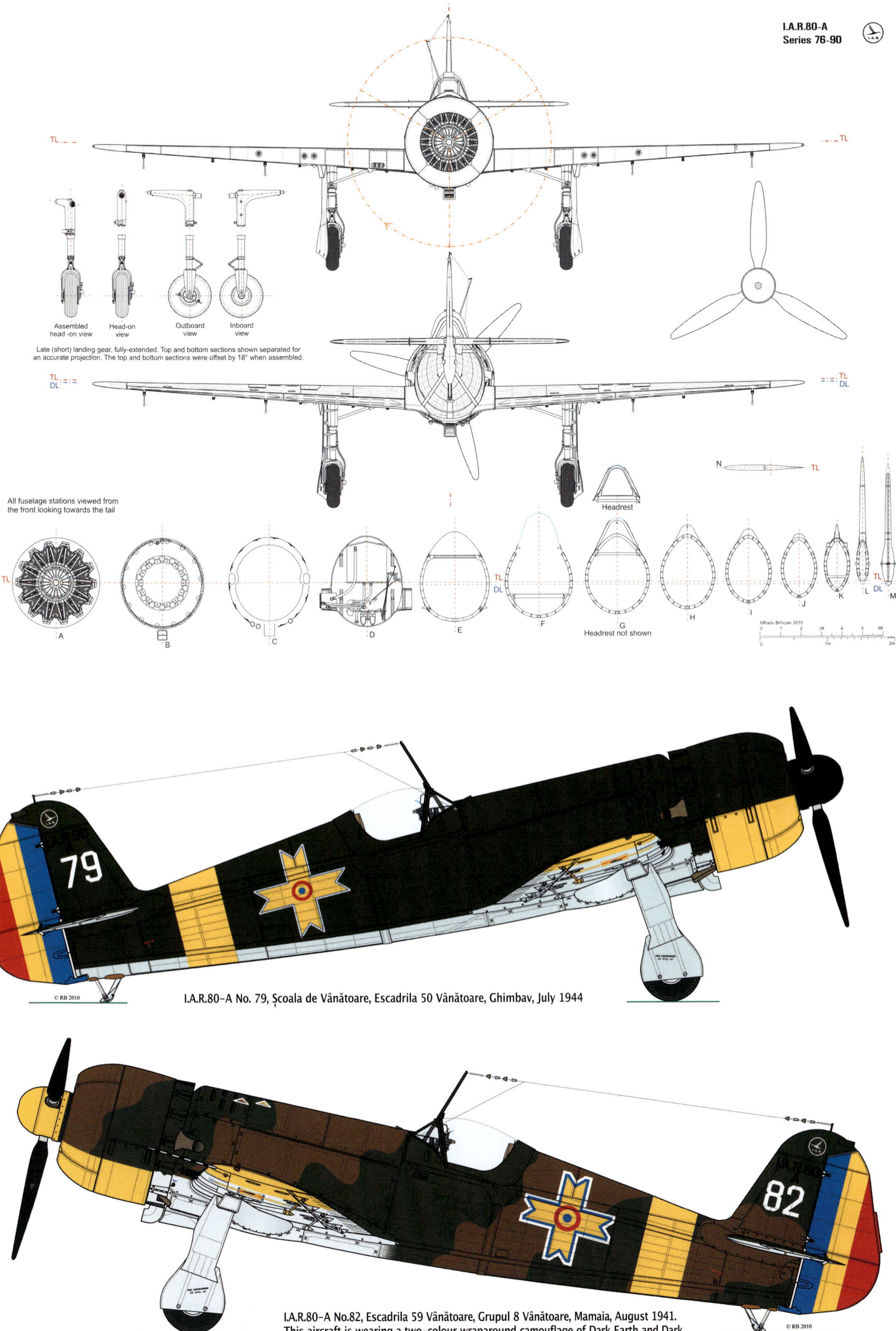

I.A.R.80-A No. 79, Şcoala de Vânătoare, Escadrila 50 Vânătoare, Ghimbav, July 1944

I.A.R.80-A No.82, Escadrila 59 Vânătoare, Grupul 8 Vânătoare, Mamaia, August 1941.
This aircraft is wearing a two-colour wraparound camouflage of Dark Earth and Dark
Green with Sky Blue wing and forward fuselage undersurfaces. Eastern Front markings

I.A.R.81 No. 91 to 105

On 2 February 1941, MAM requested I.A.R. Braşov to consider the construction of a dive bomber version of I.A.R.80. On 15 April 1942, I.A.R.80 No.1 was tested as a dive bomber equipped with the bomb launcher from a Junkers Ju 87 Stuka. The trials were carried out at the Flt.1Vt. shooting range in Codlea with excellent results. On 24 May 1941, I.A.R. Braşov fitted a bespoke central bomb launcher to I.A.R.80 No.1. This was tested with a dummy 225 kg bomb on the Codlea shooting range with excellent results. On 31 May 1941, I.A.R.80 No.54, equipped with I.A.R. bomb launchers was presented to the purchasing committee and following satisfactory results, the

bomb carrying equipment was approved. I.A.R.81 No.91 was the first in the series of I.A.R.81 BoPi dive bombers derived from the I.A.R.80-A. A central bomb rack was fitted under the fuselage, capable of carrying a payload of 250 kg. A number of oblong cut-outs were made in the panel between the wheel wells which allowed the bomb braces to pass through as well as an opening for the bomb-release lock. A parallelogram swing-arm inspired by the system used on the Junkers 87 was fitted in order to bring the bomb outside the propeller arc during the launch procedure. This device was automatically retracted with the help of a bungee cord at the end of the launch procedure. An electrical panel was fitted

I.A.R.81 No. 91 of Esc.59Vt. BoPi./Gr.6Vt./Flt.2Vt. was based in Tîrgşor, where this photograph of the I.A.R.81 flight line was taken in the autumn of 1942. The large cylindrical object on the ground is a compressed air bottle.

Photo courtesy of Jose Fernandez

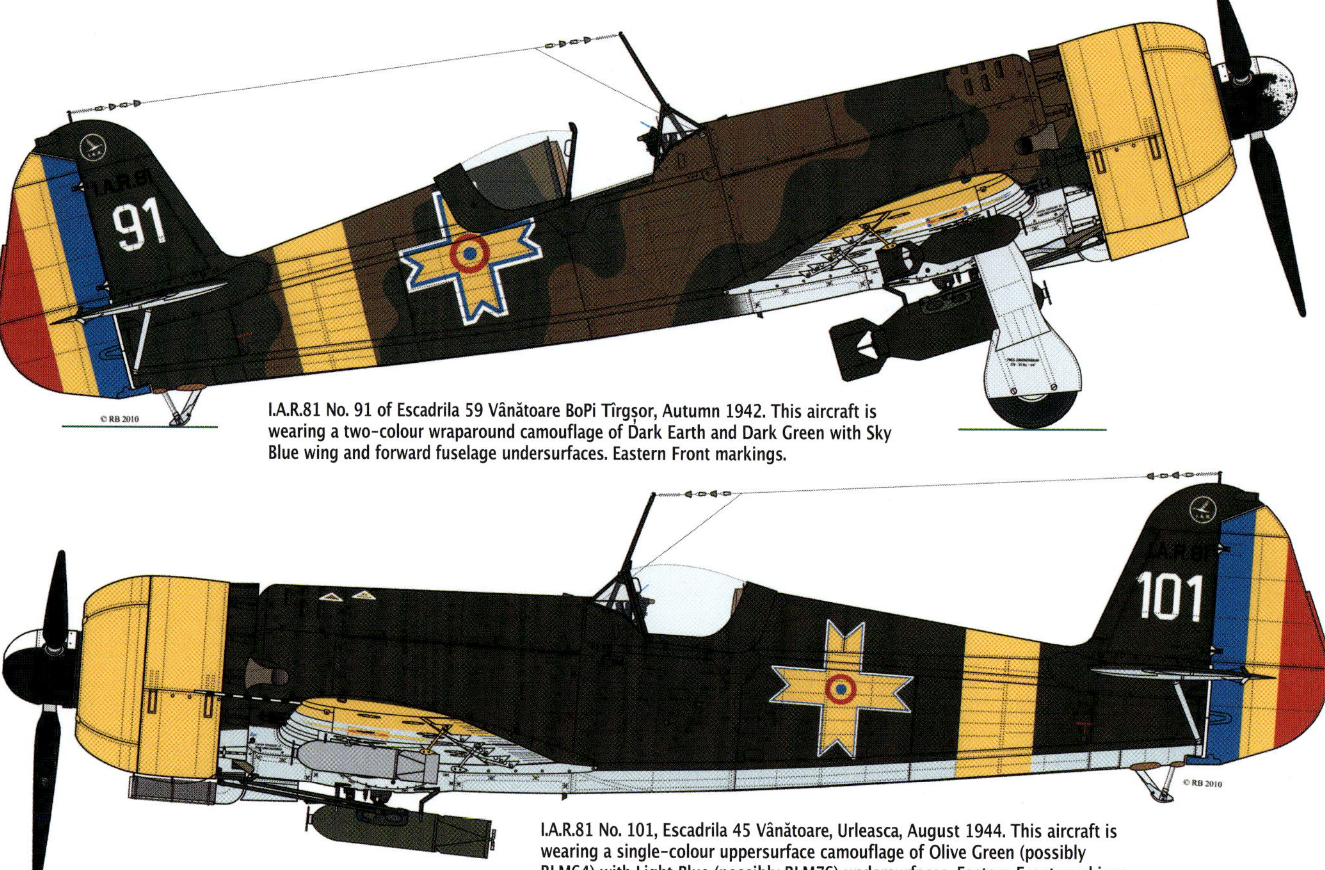

I.A.R.81 No. 91 of Escadrila 59 Vânătoare BoPi Tîrgşor, Autumn 1942. This aircraft is wearing a two-colour wraparound camouflage of Dark Earth and Dark Green with Sky Blue wing and forward fuselage undersurfaces. Eastern Front markings.

I.A.R.81 No. 101, Escadrila 45 Vânătoare, Urleasca, August 1944. This aircraft is wearing a single-colour uppersurface camouflage of Olive Green (possibly RLM64) with Light Blue (possibly RLM76) undersurfaces. Eastern Front markings

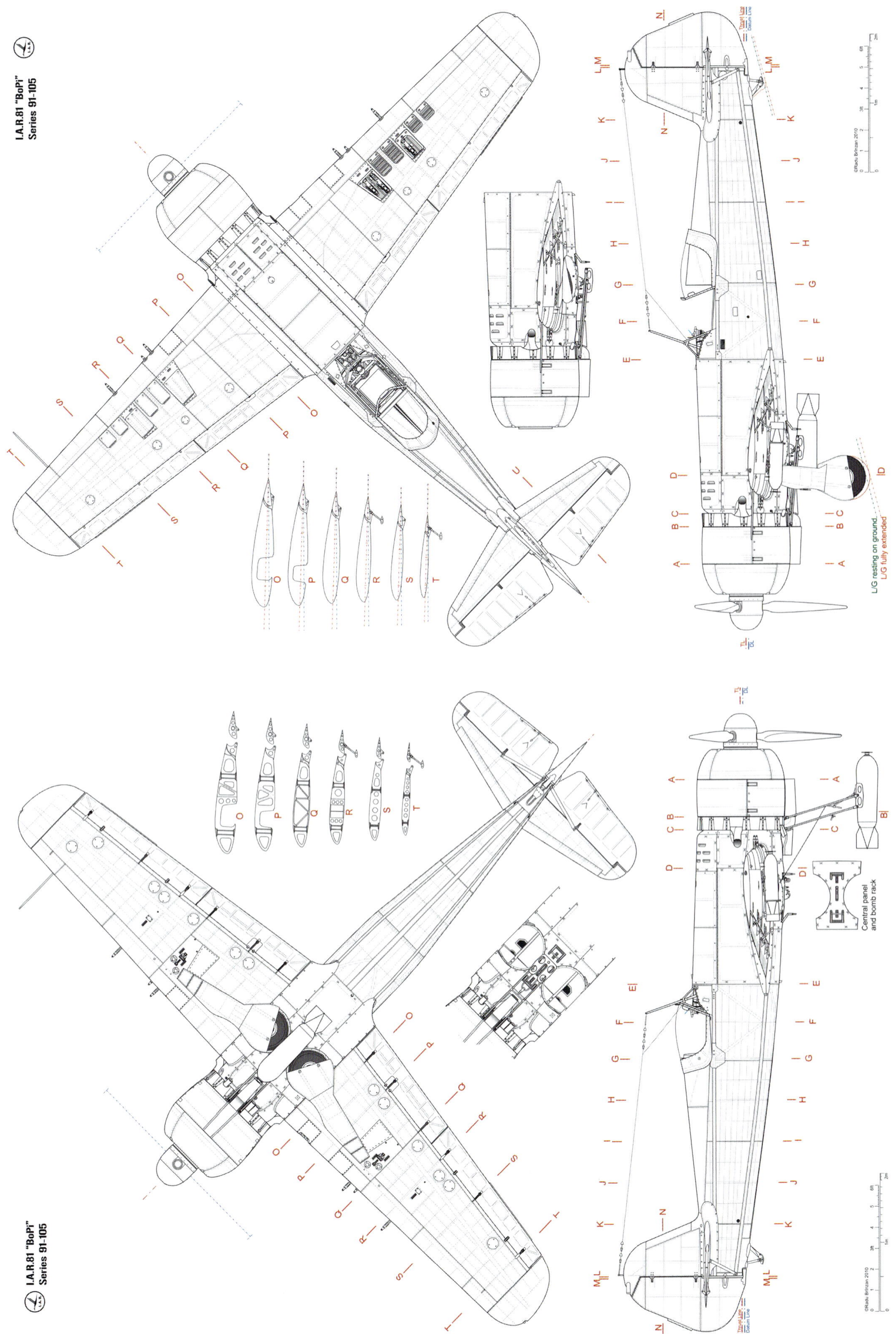
Datum Line
L/G resting on ground.
L/G fully extended
Central panel
and bomb rack
©Radu Brînzan 2010
©Radu Brînzan 2010

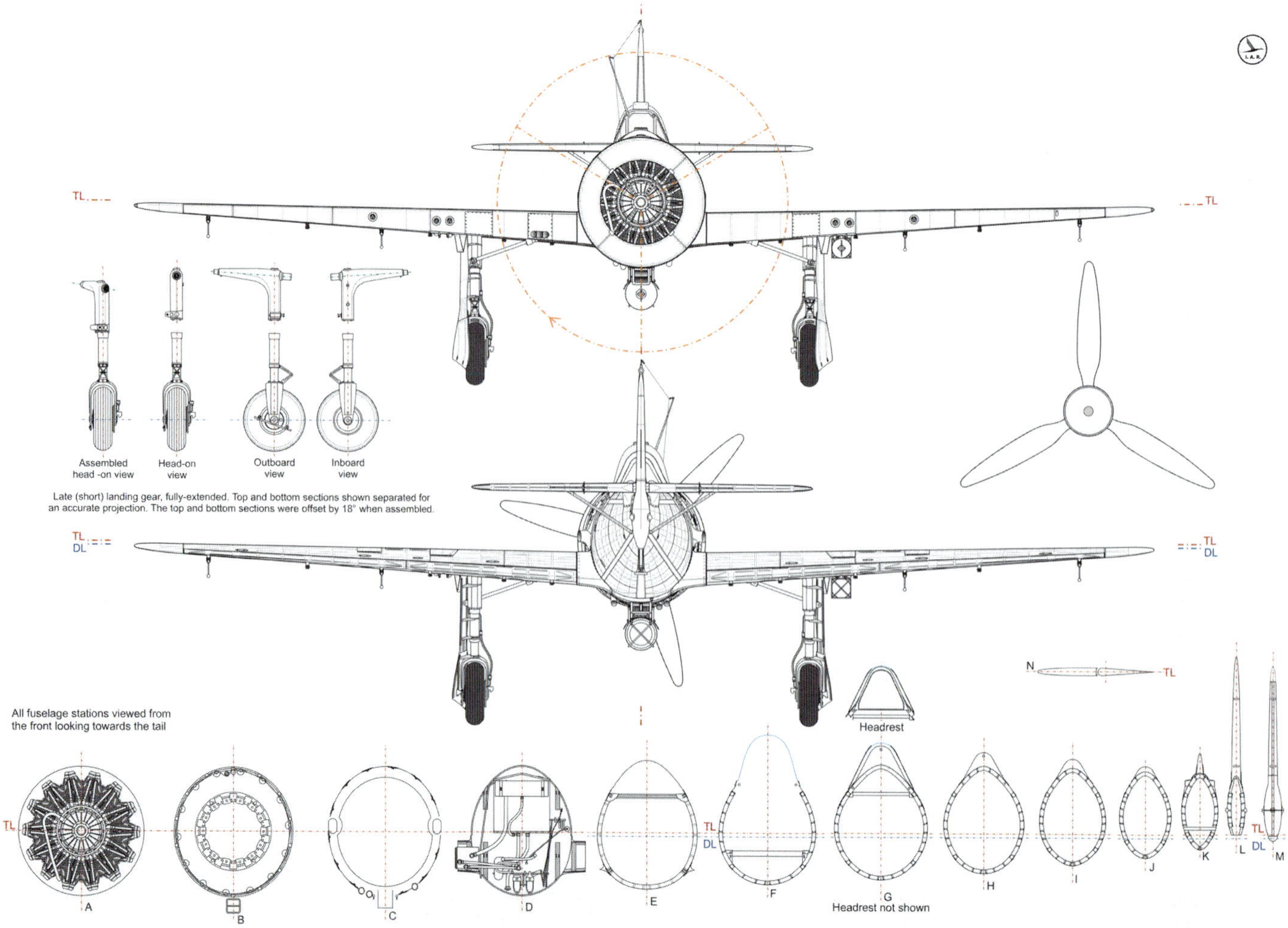

I.A.R.81 No. 101 of Gr.4Vt./Esc.45Vt. was flown on 2 August 1944 by Adj.Sef.Av.Vasile Mirilă from Urleasca in a number of bombing attacks against the Focșani Casemate. This aircraft joined this unit in June 1944 after overhaul. This picture shows the aircraft wearing a single colour olive green upper surface camouflage with a light blue fuselage underside, which was the type of camouflage usually applied to early aircraft during overhaul. Therefore, it is possible that this photo depicts the aircraft at some time between June 1944 when this aircraft was overhauled and September 1944 when the national insignia was changed. Photo courtesy of ABC collection

I.A.R.81 No.101 of Gr.4Vt./Esc.45Vt. with the engine revved-up. The aircraft is carrying one 100 kg I.A.R. bomb on the central bomb rack and two 50 kg I.A.R. bombs on the wing racks. Note the ⅓ white segment on the propeller spinner. Photo courtesy of ABC collection

Adj. Av. Eugen Jipescu poses in front of an unidentified I.A.R.81. The aircraft is armed with one licence-made 250 kg ANM-64 bomb on the central rack and two licence-made 50 kg AN-M30 bombs on the wing racks. Note the natural metal propeller blades and the ⅓ white segment on the propeller spinner. This photograph was taken with orthochromatic film making the yellow engine cowl and the national insignia appear very dark. Photo courtesy of ABC collection

inside the cockpit, which automatically retracted the flaps from the 60° dive brake position as soon as the bomb was released. An electrical switch was fitted to the control column that triggered the bomb release by using the machine gun firing lever. The tail planes were fitted with support struts. Starting with No.100, the capacity of the oil tank was increased. Starting with No.101; the installation for heating the breathing oxygen was removed. The strengthening brace riveted externally on frame 5 beginning with No.21 was riveted internally beginning with No. 95 and this modification was implemented on all subsequent airframes. Starting with No.91, the breather tube of the engine gear box was no longer connected to the exhaust collector and was extended to discharge under the fuselage, just ahead of the cockpit. The gauge of the pipe was also changed from 16 mm to 30 mm. This modification was also retrofitted to all existing airframes. As the war progressed, the armament fitted to the aircraft of this series became increasingly obsolete and many of them were upgraded with cannons and were subsequently used as fighters.

I.A.R.80-A No. 106 to 150

This series of aircraft had the same armament as that fitted to aircraft of the 91-105 series but without any bomb-carrying equipment. Starting with No.131, the shape of the headrest pillow and the head armour were modified. A strengthened windscreen with frontal and side panels made from layered glass was also introduced starting with No.131. As the war progressed, the armament fitted to the aircraft of this series became increasingly obsolete. Many were upgraded with cannons but the majority were relegated to aviation schools where they were used for training.

I.A.R.80-A No. 109, 112, 114, 111, 113 of Esc.53Vt./Gr.7Vt., I.A.R.80-A No. 149 of Esc.52Vt./Gr.9Vt. and another unidentified I.A.R.80-A are lined up on the Mamaia airfield for a training exercise in July 1942. All the aircraft in the foreground wear the 'Mounted Mickey Mouse' emblem of Esc.53Vt. None of the aircraft seem to wear the tricolour on the rudder. Reportedly, the rudders were temporarily painted red for this training exercise. Note also the stylised arrow painted on No.112. Most likely, this was a formation assembly marking denoting the aircraft of the formation leader or instructor and was applied for this exercise.
Photo courtesy of ABC Collection

This is a photo of I.A.R.80-A No. 121 after the crash in May 1946. The aircraft is still fitted with the early type of head rest and head armour. Note the late-war white markings on the fuselage and wingtips and the cockade.
Photo courtesy of Piteşti Military Archives via ABC Collection

Ground crew service I.A.R.80-A No. 133 of Esc.47Vt./Gr.9Vt. on Pipera airfield in August 1942. The aircraft was raised on jacks to service the landing gear. The hinged panel on the right side of the fuselage is open showing the internal fuselage frame and the fuel tank. The removed top fuselage panel is resting on the ground, showing the inside structure, fasteners and cooling scoops. Note the German ground power unit behind the right wing.
Photo courtesy of ABC Collection

'Mounted Mickey Mouse' was the emblem of Escadrila 53 Vânătoare. Other Disney characters such as Pluto, Pinocchio, Bambi and Donald Duck were used as emblems on many aircraft of Aeronautica Regală Română. These popular characters appeared during the war in a number of comics and children's publications such as Universul Copiilor (Children's Universe)
Drawing by Radu Brînzan

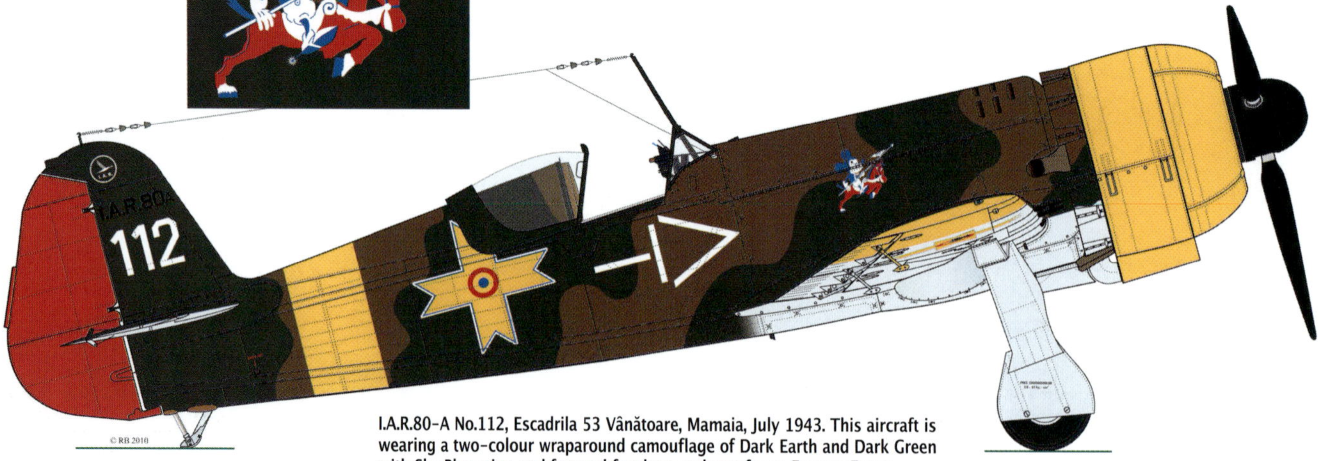

I.A.R.80-A No.112, Escadrila 53 Vânătoare, Mamaia, July 1943. This aircraft is wearing a two-colour wraparound camouflage of Dark Earth and Dark Green with Sky Blue wing and forward fuselage undersurfaces. Eastern Front markings. The rudder was temporarily painted red overall.

I.A.R.80-A No. 121 of Flotila 2 Vânătoare, Craiova, May 1946. This aircraft is wearing a single-colour uppersurface camouflage of Olive Green (possibly RLM64) with Light Blue (possibly RLM76) undersurfaces. Pro-Allied markings.

I.A.R.80-A No. 150, Escadrila 50 Vânătoare, Ghimbav, September 1944. This aircraft is wearing a single-colour uppersurface camouflage of Olive Green (possibly RLM64) with Light Blue (possibly RLM76) undersurfaces. Eastern Front markings

I.A.R.80-A No. 137, 150, 138, 179, 177 and a number of other unidentified I.A.R.80 of Esc.47Vt./Gr.9Vt. prepare for take-off from Pipera airfield in August 1942. I.A.R.80-A No. 179 was converted to an I.A.R.80 DC in 1947. Note the very low contrast between the uppersurface camouflage colours - this may be due to the use of a tinted filter on the camera lens.
Photo courtesy of ABC Collection

13 August 1942, Pipera airfield: Ground crew remove the covers from I.A.R.80-A No. 131 'Felicia' of Esc.47vt./Gr.9Vt. during a training exercise. In May 1943, this aircraft was sent to Școala de Vânătoare Flt.3Vt. in Galați. At 10.10 on 16 September 1943, this aircraft flown by Elev Av. Eugen Eftimescu hit the surface of Lake Brateș outside Galați during ground target practice. In an attempt to compensate, the pilot pulled into a steep climb to 100 metres where the plane stalled and ditched into the lake on its back. The plane was destroyed and the pilot was killed
Photo courtesy of Jose Fernandez

I.A.R.80-A No. 137 and No. 150 'Anghel' of Esc.47Vt./Gr.9Vt. prepare for take-off from Pipera airfield in August 1942. In May 1943, both aircraft were transferred to Școala de Vânătoare Flt.3Vt. Galați. Photo courtesy of ABC Collection

The four-leaf clover, a symbol of good luck, was the emblem of Escadrila 47 Vânătoare
Drawing by Teodor Liviu Moroșanu

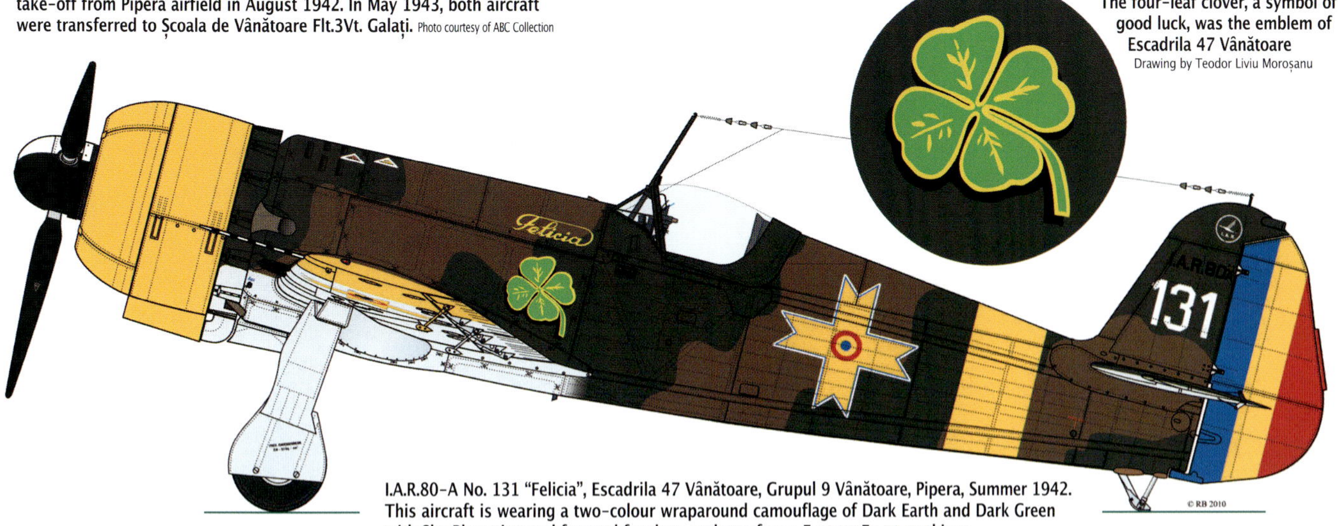

I.A.R.80-A No. 131 "Felicia", Escadrila 47 Vânătoare, Grupul 9 Vânătoare, Pipera, Summer 1942. This aircraft is wearing a two-colour wraparound camouflage of Dark Earth and Dark Green with Sky Blue wing and forward fuselage undersurfaces. Eastern Front markings.

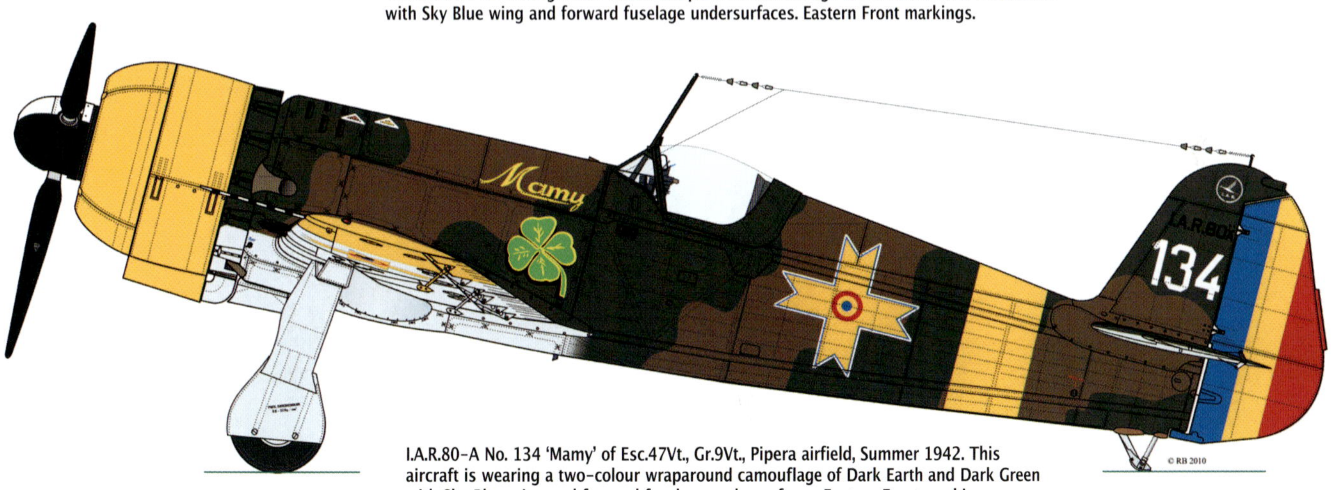

I.A.R.80-A No. 134 'Mamy' of Esc.47Vt., Gr.9Vt., Pipera airfield, Summer 1942. This aircraft is wearing a two-colour wraparound camouflage of Dark Earth and Dark Green with Sky Blue wing and forward fuselage undersurfaces. Eastern Front markings.

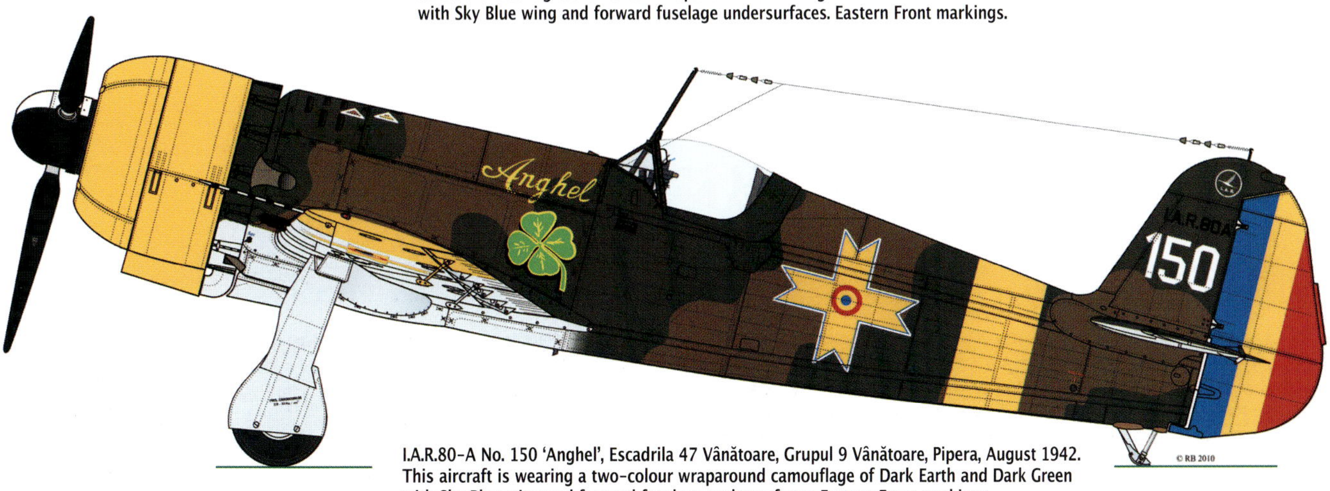

I.A.R.80-A No. 150 'Anghel', Escadrila 47 Vânătoare, Grupul 9 Vânătoare, Pipera, August 1942. This aircraft is wearing a two-colour wraparound camouflage of Dark Earth and Dark Green with Sky Blue wing and forward fuselage undersurfaces. Eastern Front markings.

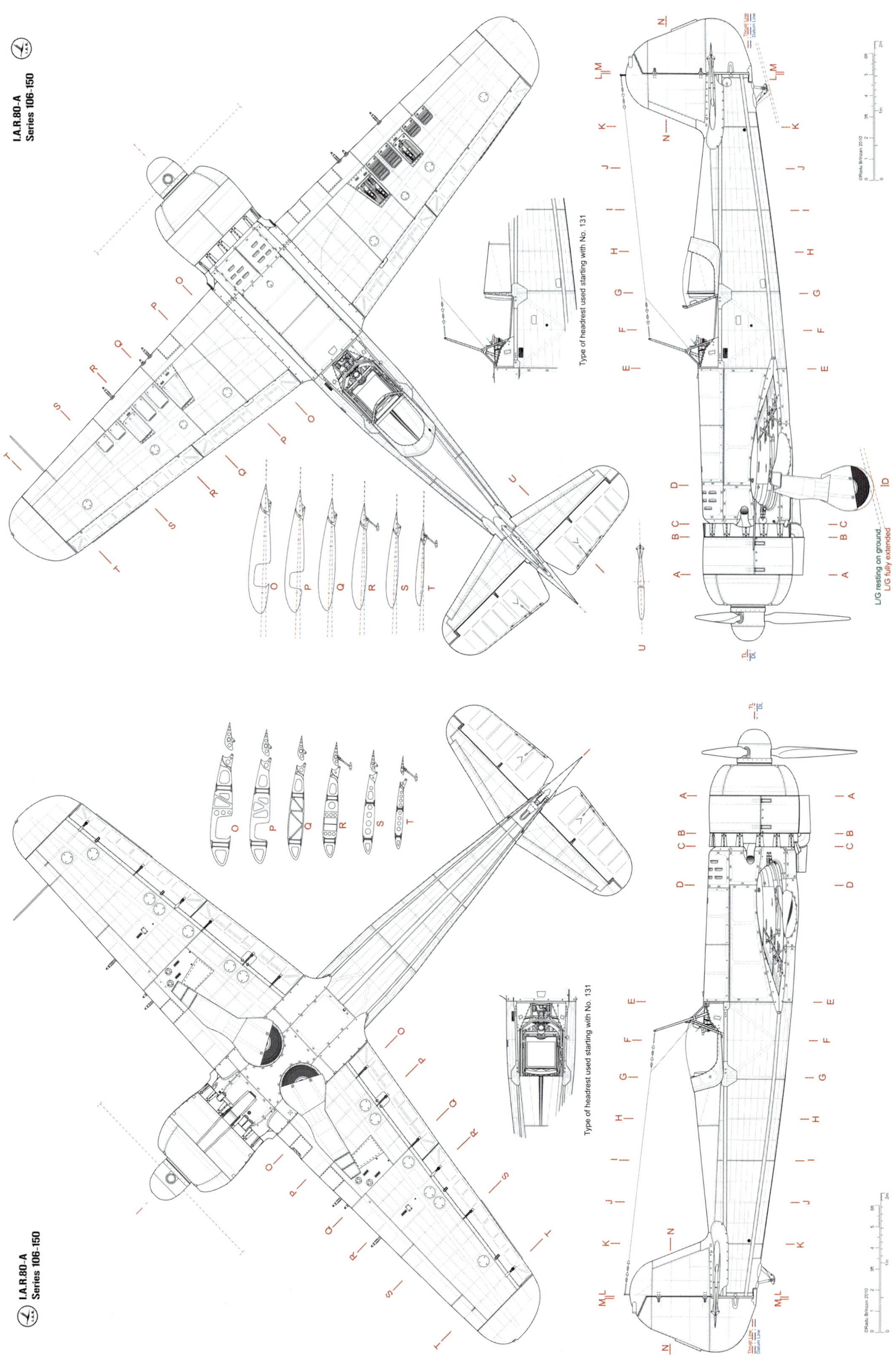
I.A.R.80-A
Series 106-150
Type of headrest used starting with No. 131
L/G resting on ground
L/G fully extended
©Radu Brinzan 2010

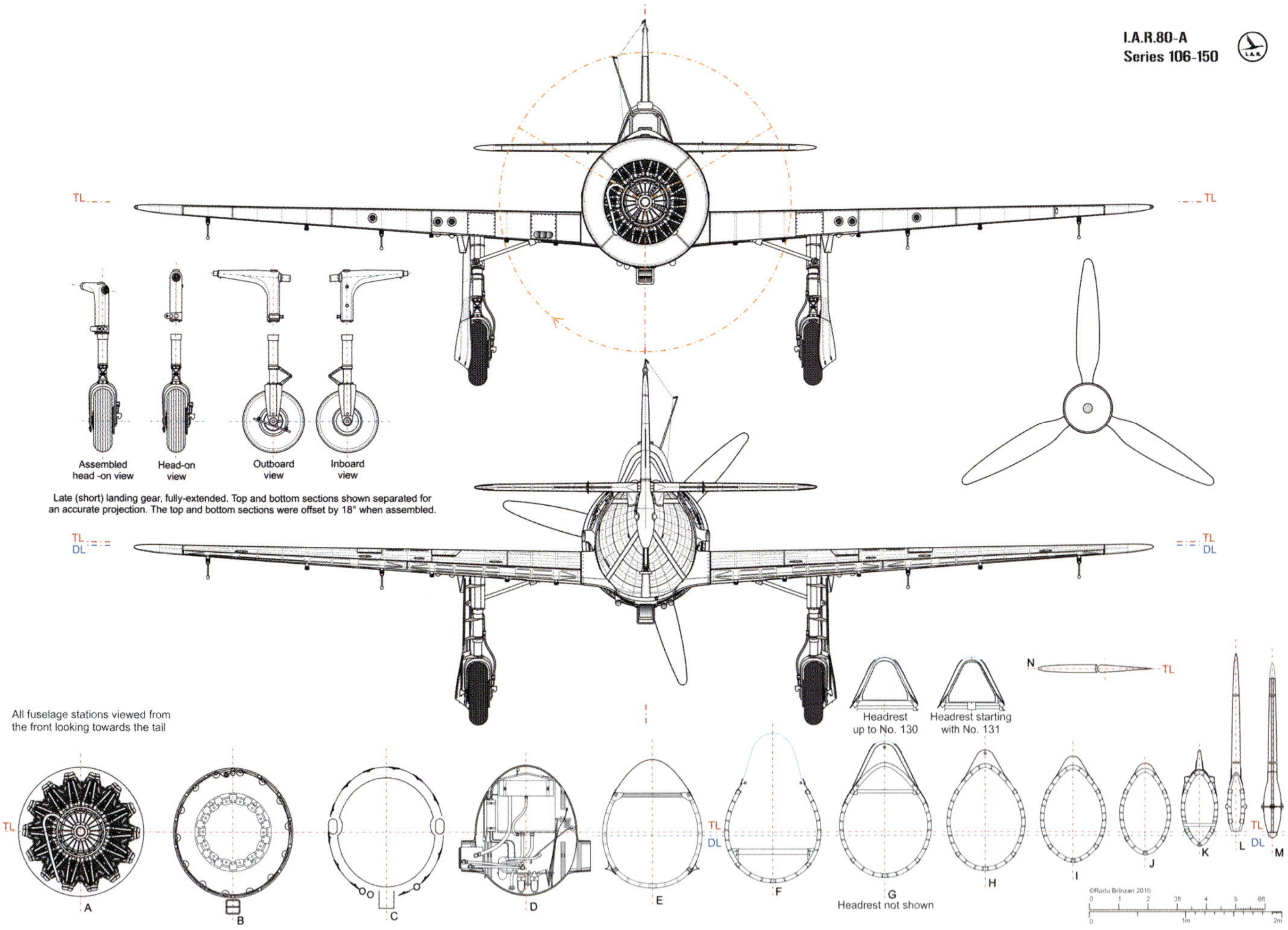

Assembled head-on view

Head-on view

Outboard view

Inboard view

Late (short) landing gear, fully-extended. Top and bottom sections shown separated for an accurate projection. The top and bottom sections were offset by 18° when assembled.

All fuselage stations viewed from the front looking towards the tail

Headrest up to No. 130

Headrest starting with No. 131

Headrest not shown

©Radu Brînzan 2010

At 11.40 on 26 April 1942, I.A.R.80-A No. 145 of Esc.47Vt./Gr.9Vt. flown by Adj.Av. Ioan Tura stalled while landing and belly-landed 10 metres from the edge of Pipera airfield. The aircraft was repaired by I.A.R. Brașov after which it was sent to Școala de Vânătoare Flt.3Vt. Galați. In May 1944, this aircraft was transferred to Esc.44Vt. in Ghimbav. On 6 July 1944, this aircraft flown by Adj.Av. Ion Stoican collided in mid-air with I.A.R.80 No. 68 flown by Elev Serg. Ioan Stoica. I.A.R.80-A No. 145 and the pilot were lost. I.A.R.80-A No. 68 managed to land, but it was later scrapped.

Photo courtesy of Pitești Military Archives via ABC Collection

I.A.R.80-A No. 142 and 146 of Esc.47Vt./Gr.9Vt. and a number of unidentified I.A.R.80 are pictured on Pipera airfield in the summer of 1942.

Photo courtesy of ABC Collection

I.A.R.-81 No. 151 to 175

I.A.R.81 No.s 167, 166, 100, 239, 232 and 98 of Esc.59Vt.
BoPi/Gr.6Vt./Flt.2Vt. were photographed on Tîrgşor airfield in the
autumn of 1942. At this time, I.A.R.81 No. 167 was fitted with an
accelerometer and used to test the aircraft in near-vertical dives
Photo courtesy of Jose Fernandez

Aircraft of this series had the same armament as that fitted to aircraft of the 91 to 105 series with all the airframe modifications introduced up to No.150. A new type of flap strengthened by the addition of a long chrome-molybdenum sleeve that the control lever was welded to was introduced starting with this series. These changes were only internal. The tail planes were fitted with support struts. As the war progressed, the armament fitted to the aircraft of this series became increasingly obsolete and many of them were upgraded with cannons and were used as fighters.

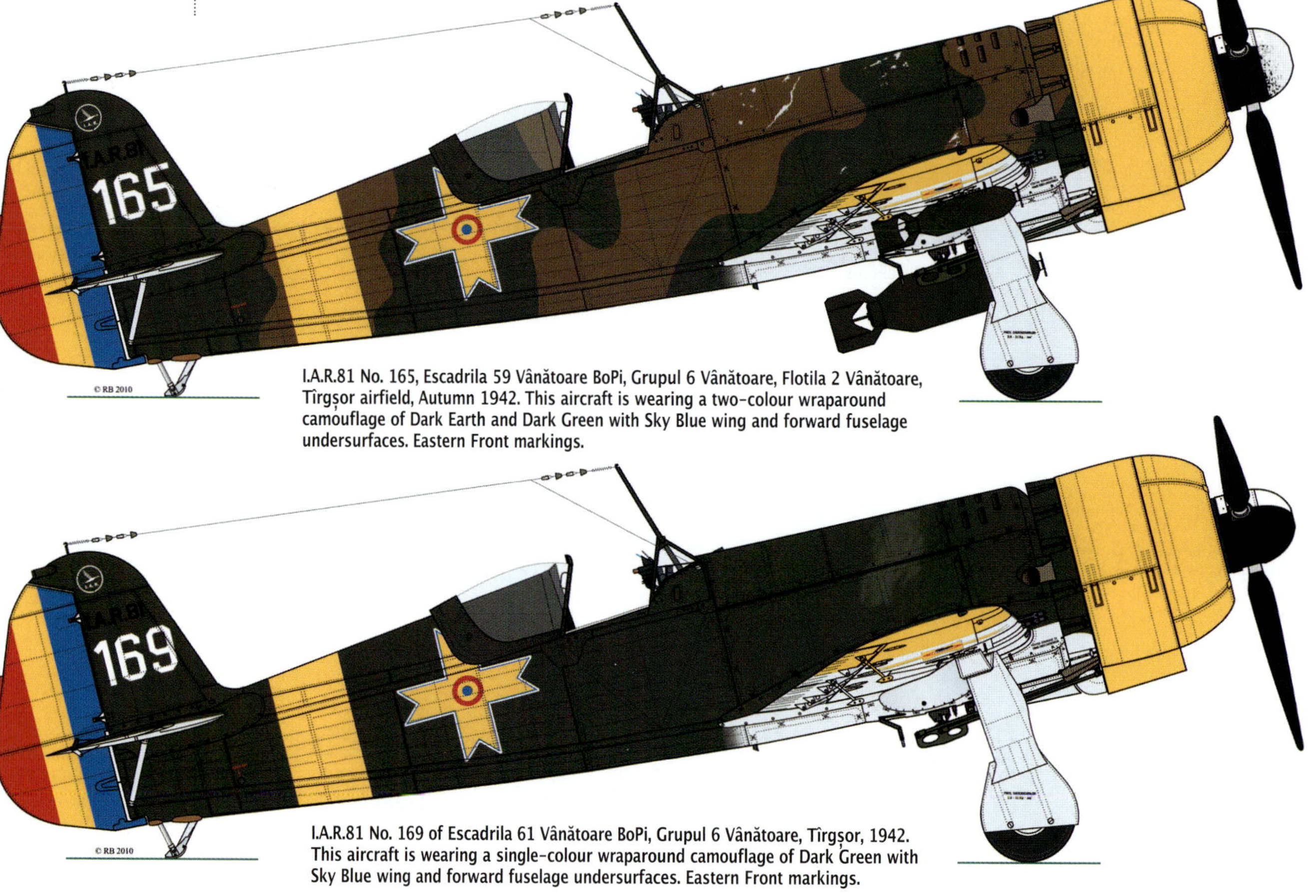

I.A.R.81 No. 165, Escadrila 59 Vânătoare BoPi, Grupul 6 Vânătoare, Flotila 2 Vânătoare,
Tîrgşor airfield, Autumn 1942. This aircraft is wearing a two-colour wraparound
camouflage of Dark Earth and Dark Green with Sky Blue wing and forward fuselage
undersurfaces. Eastern Front markings.

I.A.R.81 No. 169 of Escadrila 61 Vânătoare BoPi, Grupul 6 Vânătoare, Tîrgşor, 1942.
This aircraft is wearing a single-colour wraparound camouflage of Dark Green with
Sky Blue wing and forward fuselage undersurfaces. Eastern Front markings.

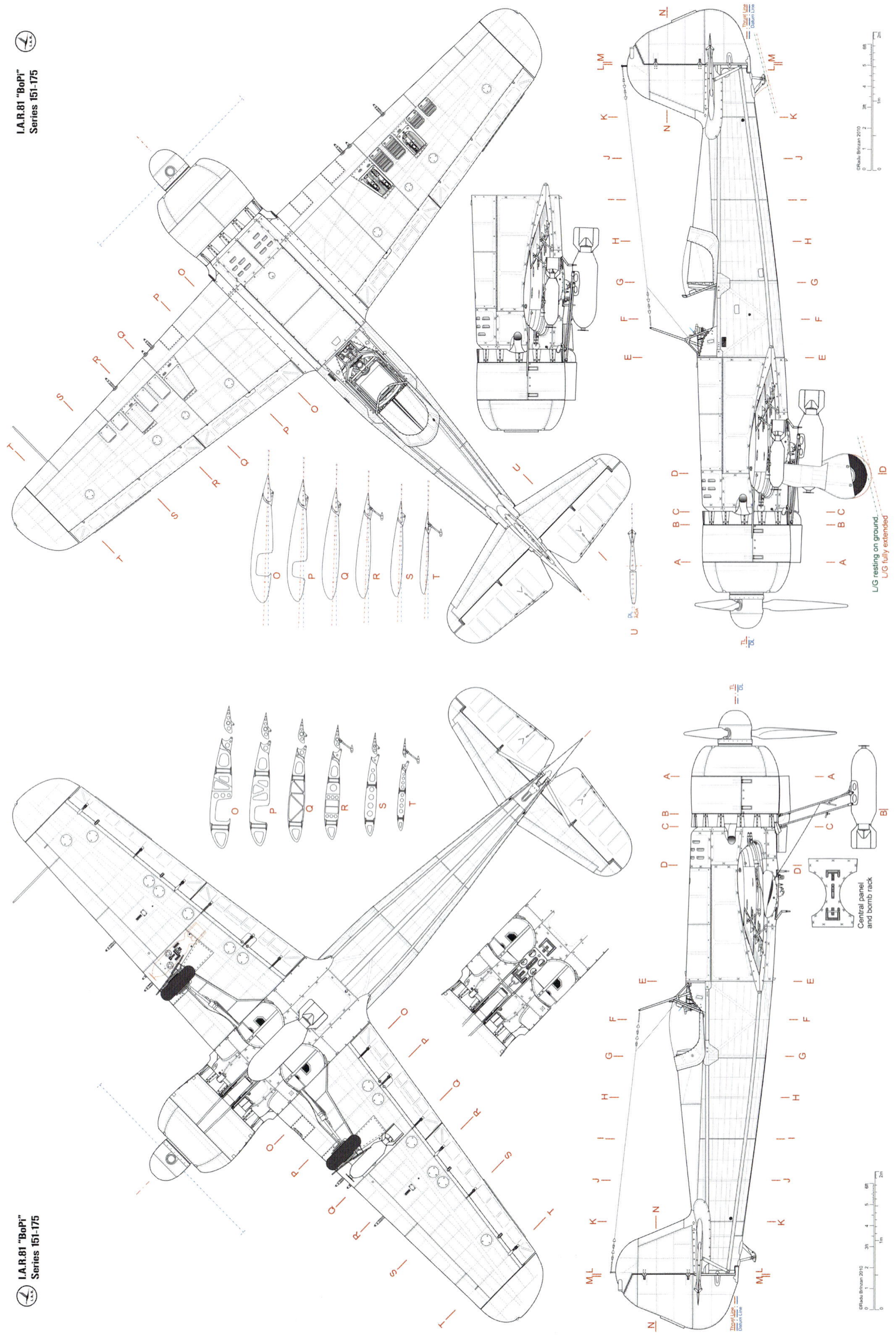
L/G resting on ground
L/G fully extended
Central panel
and bomb rack
©Radu Brinzan 2010
©Radu Brinzan 2010

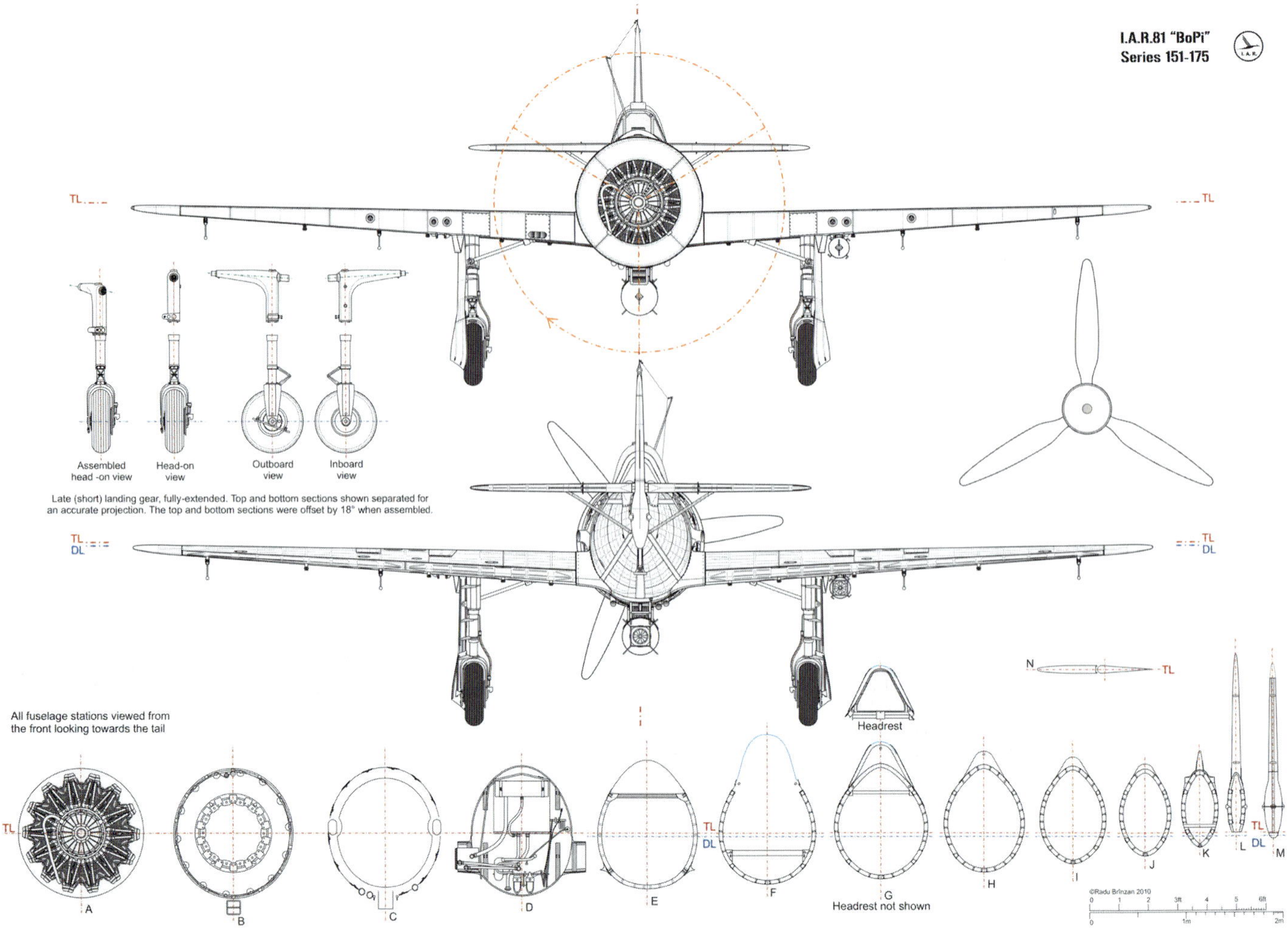
Assembled head -on view
Head-on view
Outboard view
Inboard view
Late (short) landing gear, fully-extended. Top and bottom sections shown separated for an accurate projection. The top and bottom sections were offset by 18° when assembled.
TL
DL
N
Headrest
All fuselage stations viewed from the front looking towards the tail
A
B
C
D
E
F
G
Headrest not shown
H
I
J
K
L
M
©Radu Brînzan 2010

169

This photograph of I.A.R.81 No. 169 of Esc.61Vt. BoPi/Gr.6Vt. was taken on Tîrgșor airfield in the autumn of 1942 when it was flown by Adj.Ș.Av. Vasile Mirilă in training missions with practice and real bombs. This aircraft was later used on the Eastern Front from Tusov and Morozovskaya airfields in late-1942. In early-1943, this aircraft was returned for repairs to ASAM Cotroceni from where it was sent to I.A.R. Brașov in November of the same year to be upgraded with Mauser cannons. Note the single colour wrap-around camouflage scheme and the ⅓ white segment on the propeller spinner.
Photo courtesy of ABC Collection

I.A.R.81
91
165
167
166

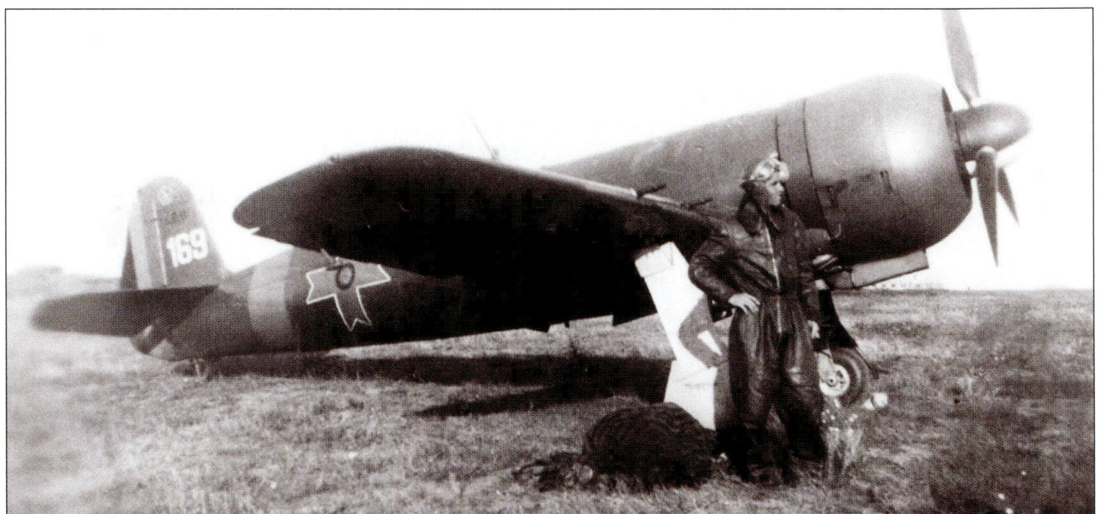

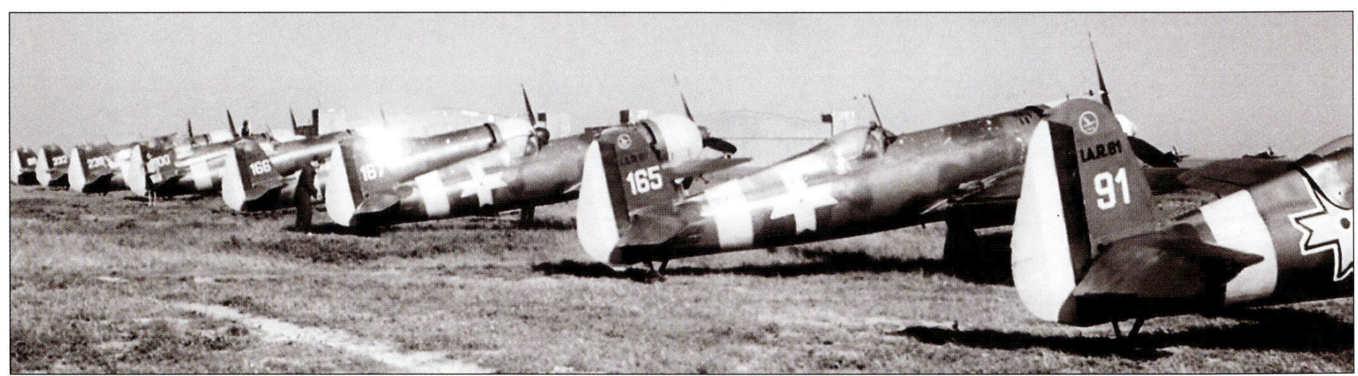

I.A.R.81 No. 91, 165, 167, 166, 100, 239, 232 and 98 of Esc.59Vt. BoPi/Gr.6Vt./Flt.2Vt. were photographed in Tîrgșor in the autumn of 1942.
Photo courtesy of Jose Fernandez

I.A.R.80-A No. 176 to 180

I nitially ordered as I.A.R.80-B, the aircraft of this very short series were built as I.A.R.80-A due to shortage of 13.2 mm machine guns. Aircraft of this series had the same armament as that fitted to aircraft of the 131 to 150 series with all the airframe modifications introduced up to No.170. Aircraft No. 176 was used to test the armament configuration of 4 x 7.92 mm FN Browning and 2 x 13.2 mm FN Browning machine guns. As the war progressed, the armament fitted to the aircraft of this series became increasingly obsolete. Many aircraft of this series were relegated to aviation schools where they were used for training.

After an accident in Galați in 1943, I.A.R.80-A No. 178 was sent to I.A.R. Brașov where it was repaired and returned to Școala de Vânătoare Flt.3Vt. in Galați in November 1943. In April 1944, it was transferred to Esc.50Vt./Gr.3Vt. in Ghimbav where it was damaged during the bombing by USAF on 14 June 1944. This photograph shows the damage sustained by the aircraft in that attack. The aircraft was repaired and survived until 1952 when it was scrapped in Focșani.
Photo courtesy of Pitești Military Archives via ABC Collection

I.A.R.80-A No. 179, Școala de Vânătoare, Flotila 3 Vânătoare Galați, June 1943. This aircraft is wearing a two-colour wraparound camouflage of Dark Earth and Dark Green with Sky Blue wing and forward fuselage undersurfaces. Eastern Front markings.

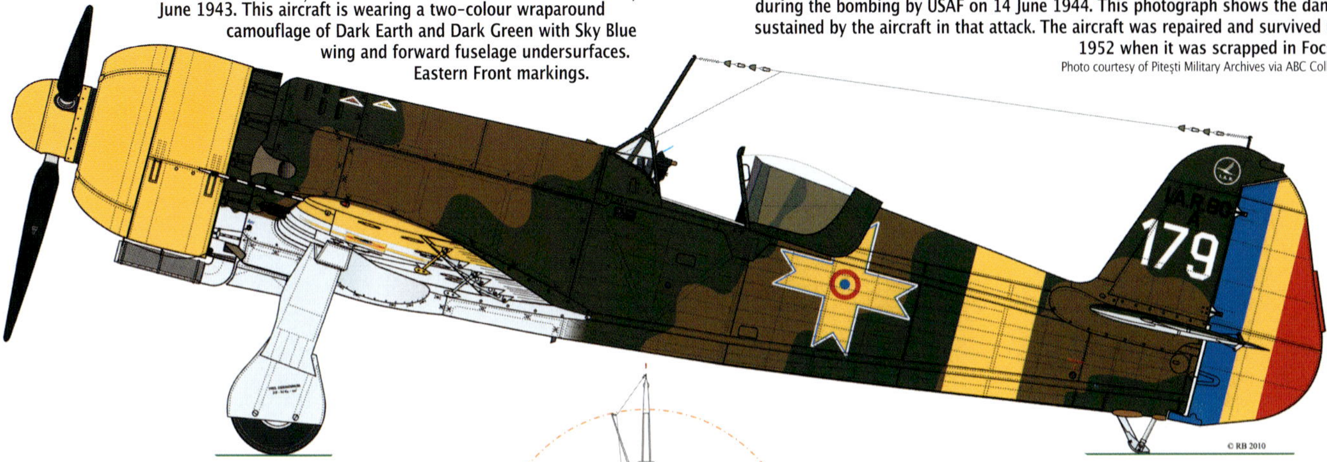

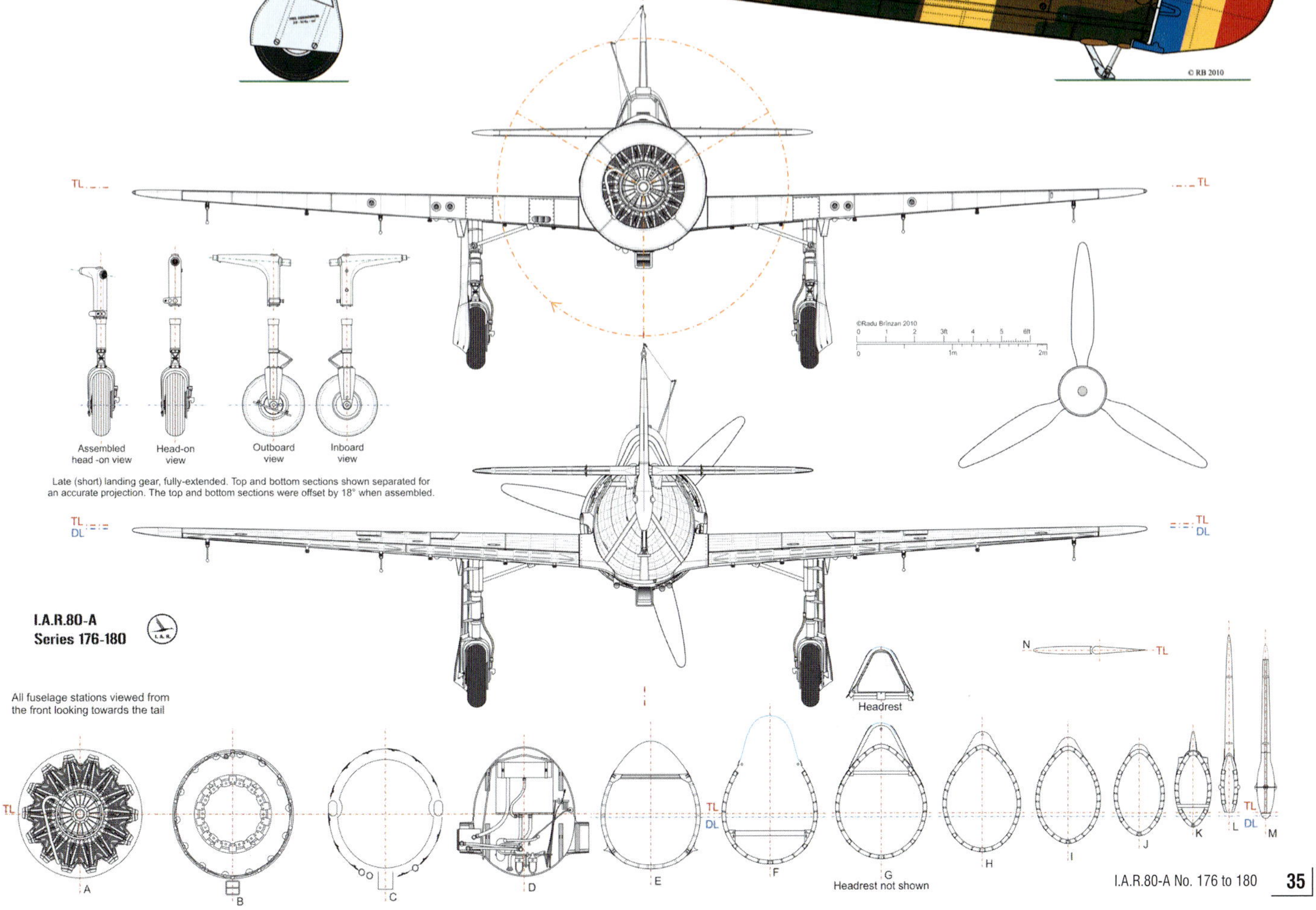

Assembled head-on view

Head-on view

Outboard view

Inboard view

Late (short) landing gear, fully-extended. Top and bottom sections shown separated for an accurate projection. The top and bottom sections were offset by 18° when assembled.

I.A.R.80-A Series 176-180

All fuselage stations viewed from the front looking towards the tail

Headrest

Headrest not shown

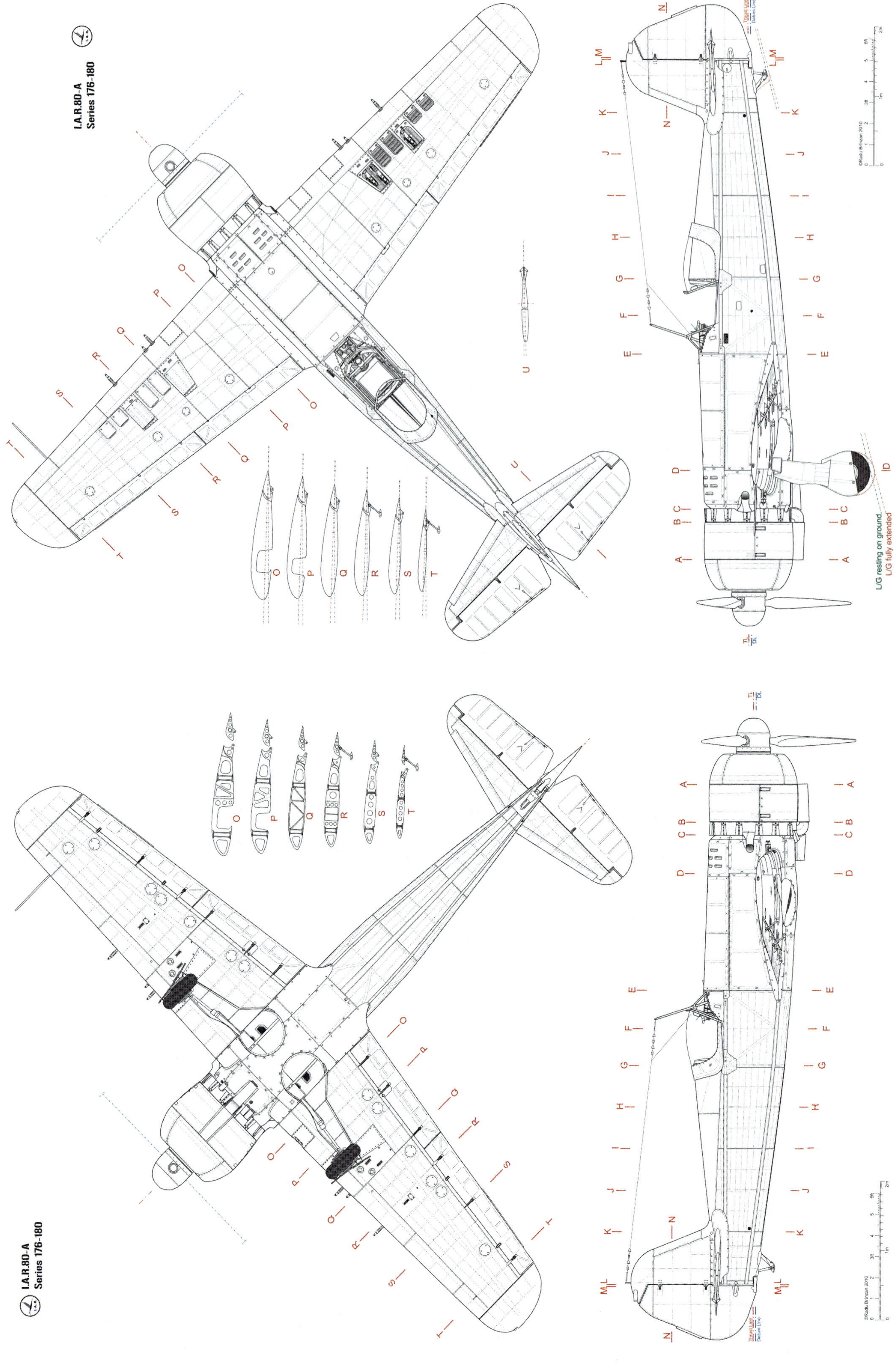

I.A.R.80-A
Series 176-180
L/G resting on ground
L/G fully extended
©Radu Brinzan 2010
Thrust Line
Datum Line

I.A.R.80-B No. 181 to 200

This series was fitted with a new heavier armament of 2 x 13.2 mm FN Browning and 4 x 7.92mm FN Browning machine guns. The ammunition carried was 1,600 x 7.92 mm rounds and 300 x 13.20mm rounds. The outboard 7.92mm guns were located in the same position as the outboard guns of the earlier-type six-gun I.A.R.80, but the middle 7.92mm guns were relocated slightly outboard in comparison to earlier-type aircraft in order to provide clearance for the 13.20mm gun. All guns were fired electrically. The 13.20mm guns were charged pneumatically, whereas the 7.92mm guns were charged manually on the ground. The grip of the control stick was replaced with an oval 'spade grip' with three electrical switch buttons, one for the 7.92mm guns, one for the 13.20mm guns and one for the radio. The new grip also had a lever controlling the brakes. A rubber gasket was fitted around the head armour to provide a certain amount of draft exclusion for the cockpit. This type of headrest was used on all further aircraft and retrofitted to some earlier-type airframes.

Between August and December 1942, I.A.R.80-B No. 200 of Esc.41Vt./Flt.2Vt. was flown by Lt.Av. Traian Gavriliu on the Stalingrad front from the Tusov and Morozovskaya airfields. The unusually extensive wear on the paintwork was caused by the harsh weather conditions and the rough airfields that the aircraft operated from. Note also the dark green engine cowl. On 10 September 1942, Lt.Av. Traian Gavriliu claimed one Soviet Yak shot down while flying this aircraft. In December 1942, this aircraft was sent back to ASAM Cotroceni for overhaul and after repairs it was transferred to Esc.64Vt./Gr1.Vt. in Roșiorii de Vede where it saw some combat action against USAF aircraft in the spring of 1944 without any victory claims. Photo courtesy of ABC collection

A group of pilots and ground-crewmen pose in front of an unidentified I.A.R.80-B.
Photo courtesy of ABC collection

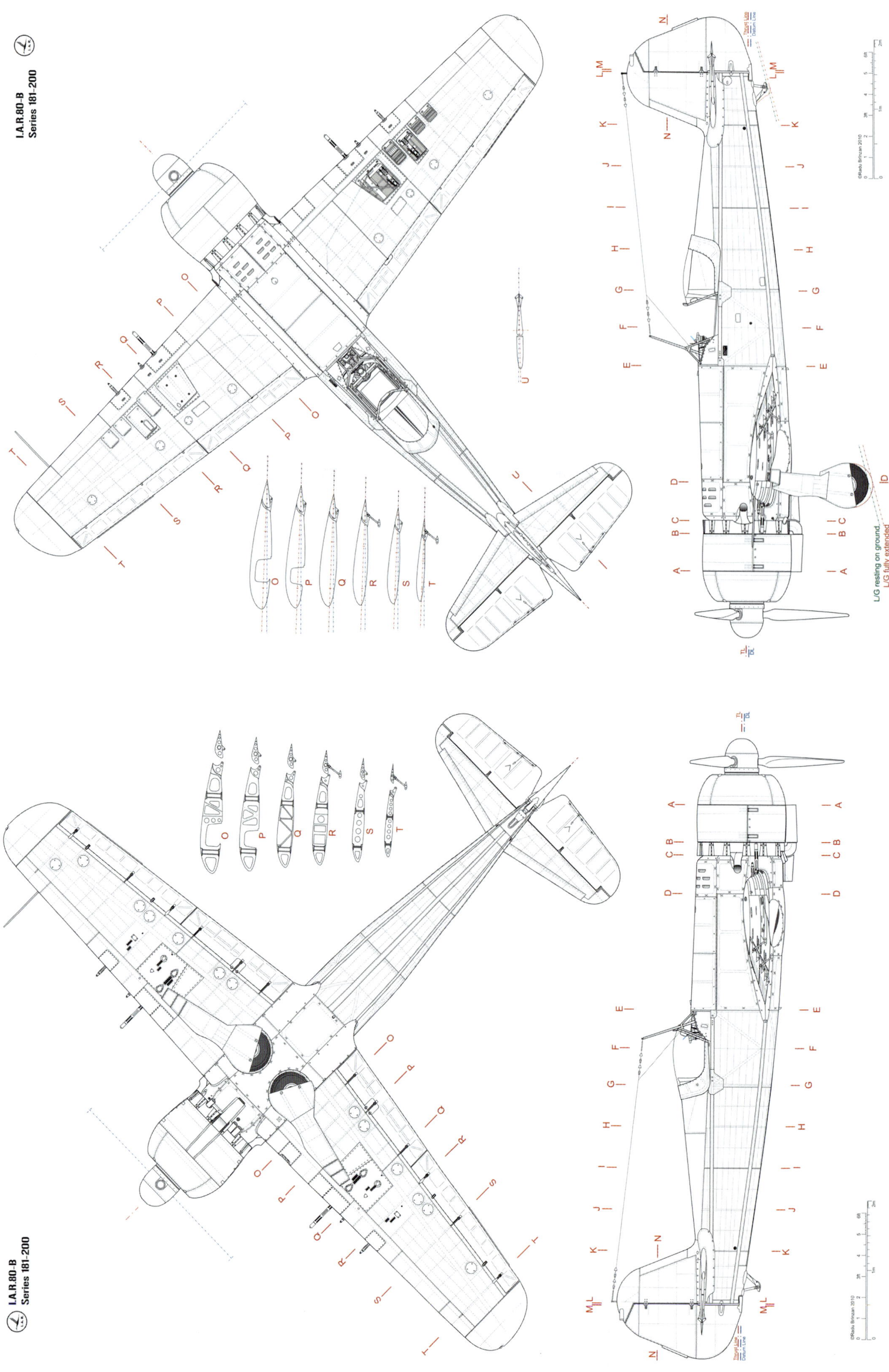

I.A.R.80-B
Series 181-200
I.A.R.80-B
Series 181-200
L/G resting on ground
L/G fully extended
©Radu Brinzan 2010
©Radu Brinzan 2010

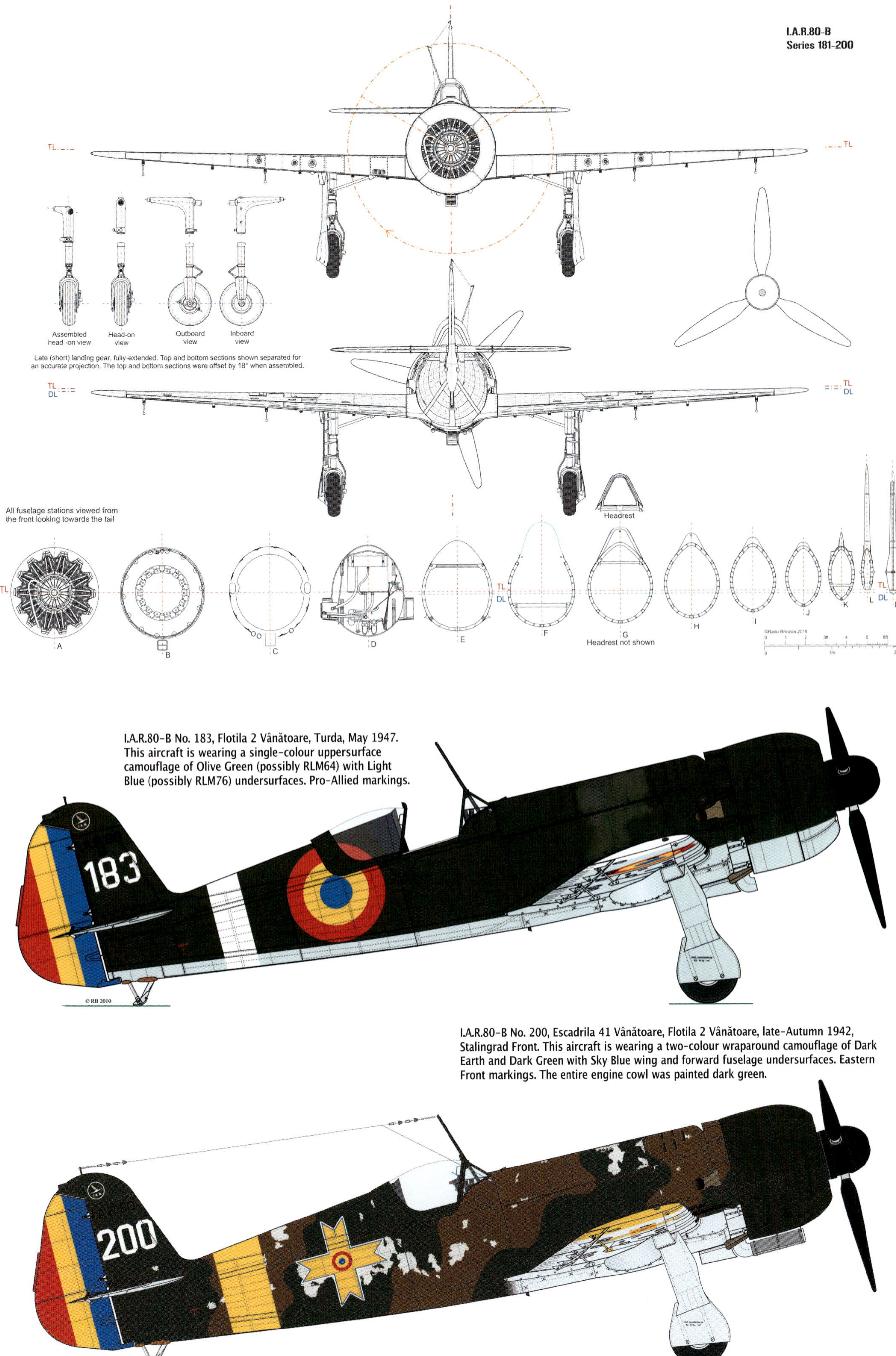

I.A.R.80-B No. 183, Flotila 2 Vânătoare, Turda, May 1947.
This aircraft is wearing a single-colour uppersurface
camouflage of Olive Green (possibly RLM64) with Light
Blue (possibly RLM76) undersurfaces. Pro-Allied markings.

I.A.R.80-B No. 200, Escadrila 41 Vânătoare, Flotila 2 Vânătoare, late-Autumn 1942,
Stalingrad Front. This aircraft is wearing a two-colour wraparound camouflage of Dark
Earth and Dark Green with Sky Blue wing and forward fuselage undersurfaces. Eastern
Front markings. The entire engine cowl was painted dark green.

I.A.R.80-B No. 201 to 211

The aircraft of the 201 to 211 series were initially planned as I.A.R.81-A dive bombers with heavy machine guns, but they were built as fighters with the same armament as that fitted to aircraft of the 181-200 series with all the airframe modifications introduced up to No.200. Starting with No. 201, the fuselage was extended by 70 mm ahead of the firewall. Also starting with this series; the aircraft were fitted with under-wing racks and the necessary plumbing for carrying drop tanks. A second breather tube for the engine gear box was added on the left side of the engine, symmetrical with the tube on the right side but shorter. This was connected to the left exhaust collector.

I.A.R.80-B No. 201 was photographed on Mizil airfield where it was based between June and August 1942 as part of Esc.60Vt./Gr.8Vt. At 13.58 on 21 August 1942, this aircraft flown by Slt.Av. Mihai Slăvescu crash-landed near the Mărculești railway station due to a navigation error. The aircraft was repaired by ASAM Cotroceni and in January 1944 it was sent to Esc.63Vt./Flt.2Vt. based in Roșiorii de Vede. On 5 May 1944, I.A.R.80-B No. 201 flown by Adj.Av. Virgil Anghelescu was shot down during combat with USAF aircraft and belly-landed near Glovacioc. The aircraft was repaired and flew again.

Photo courtesy of ABC Collection

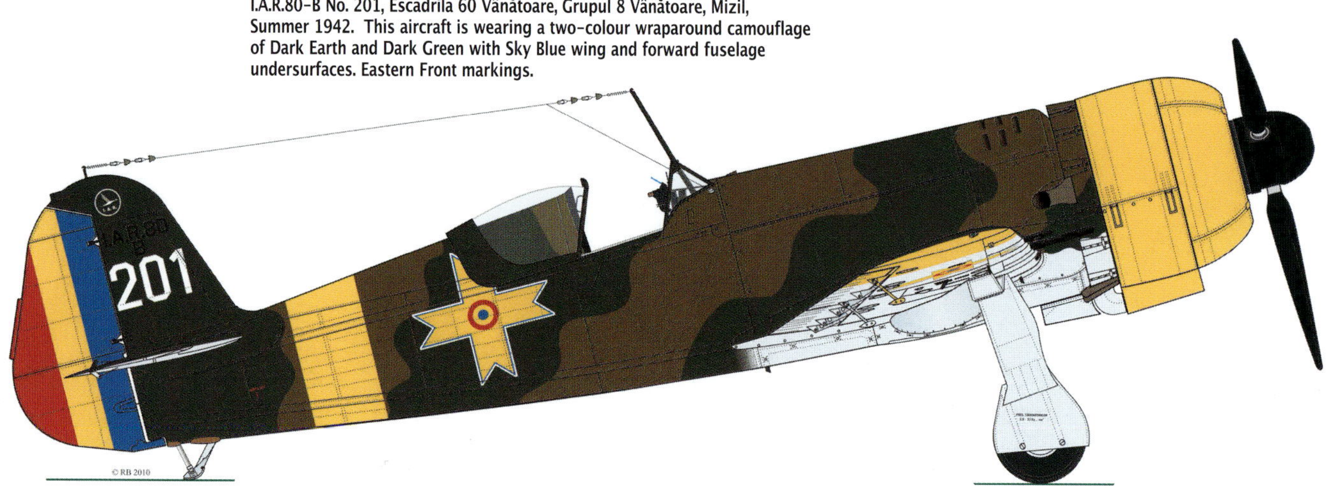

I.A.R.80-B No. 201, Escadrila 60 Vânătoare, Grupul 8 Vânătoare, Mizil, Summer 1942. This aircraft is wearing a two-colour wraparound camouflage of Dark Earth and Dark Green with Sky Blue wing and forward fuselage undersurfaces. Eastern Front markings.

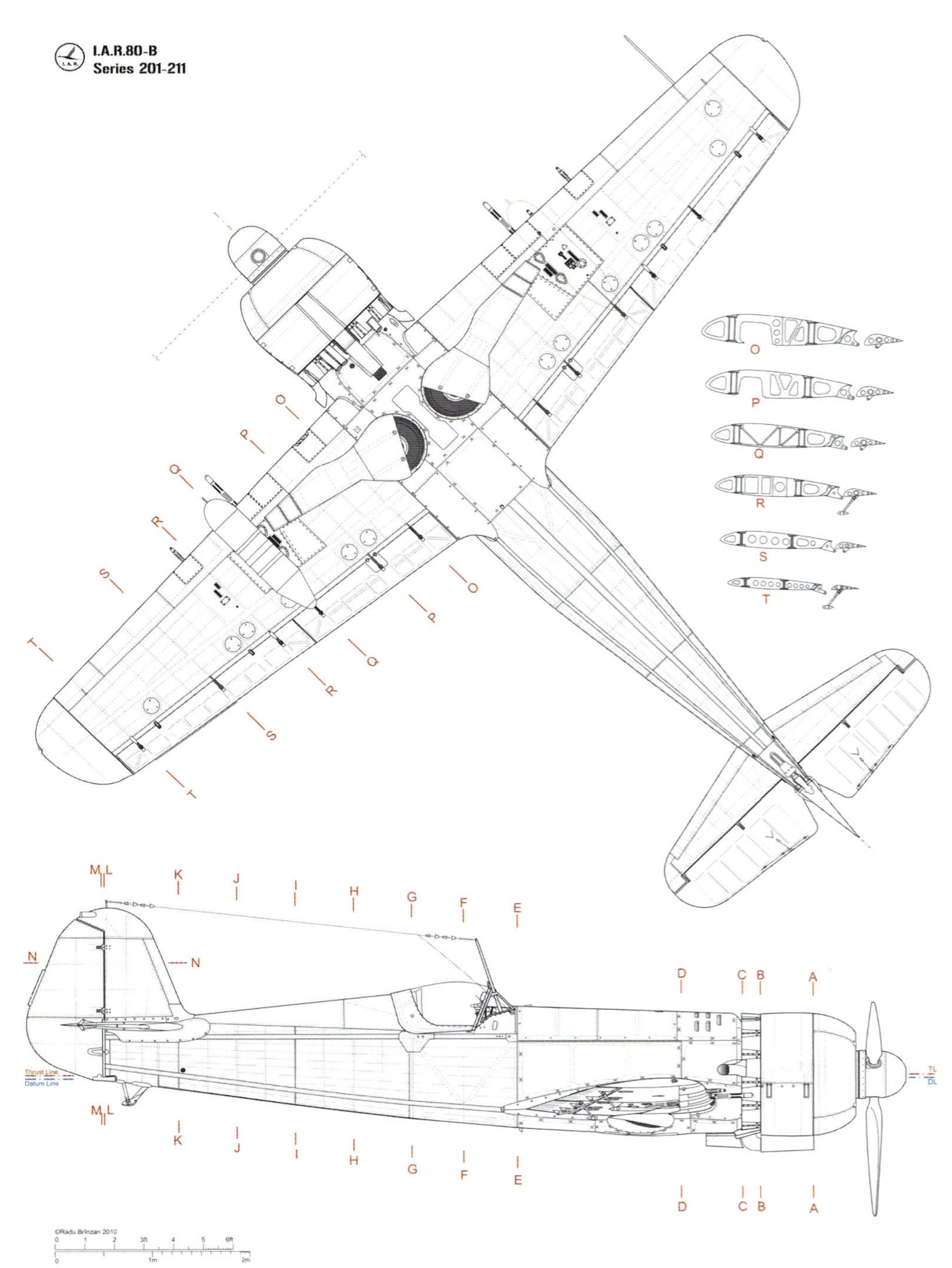

I.A.R.80-B
Series 201-211
I.A.R.80-B
Series 201-211
Thrust Line
Datum Line
L/G resting on ground.
L/G fully extended.
©Radu Brinzan 2010
©Radu Brinzan 2010

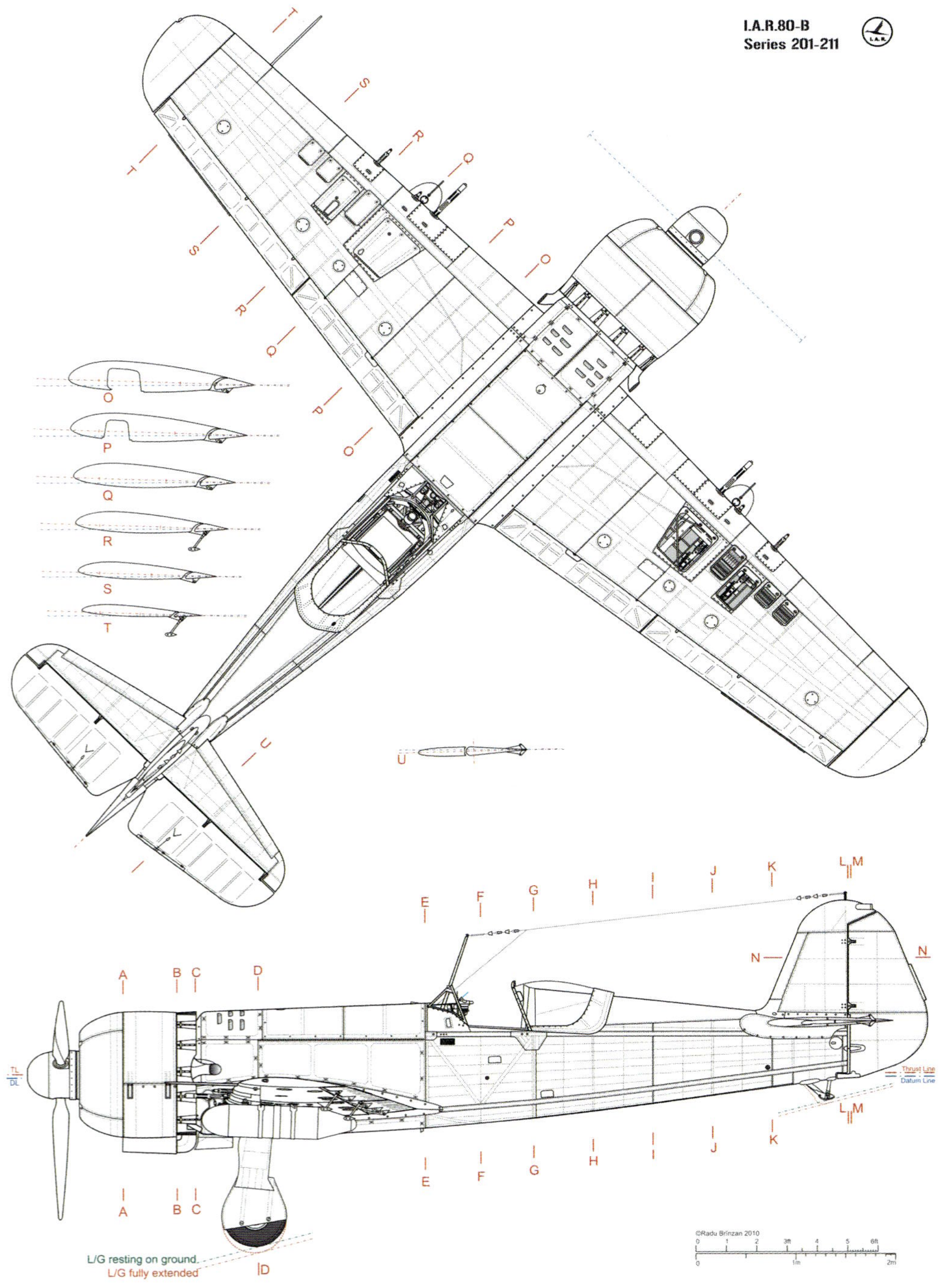

Another view of I.A.R.80-B No. 203 after the crash in August 1950. It appears that the spinner still wears a spiral, an Axis recognition marking, which should have been removed from all aircraft wearing late-war markings. The spinner may be a replacement from storage that was not repainted. Also, the nose of this aircraft is fitted with the type of side cooling gills typical for the aircraft of the series 251 to 290 as well as tail plane support struts which is a further indication that this aircraft was repaired using spares from other aircraft
Courtesy of Pitești Military Archives via ABC Collection

On 25 February 1944, I.A.R.80-B No. 211 of Esc.64Vt./Gr.1Vt. flown by Adj.Maj.Av. Andrei Ciobănaș was caught in snow during take-off and overturned on Roșiorii de Vede airfield. The aircraft was repaired and returned to flight status. On 9 September 1944, while part of Esc.44Vt., this aircraft was set on fire and destroyed by German troops on Ghimbav airfield. This photograph provides an excellent view of the armament bulges and spent ammunition chutes associated with the heavy machine gun wing. Note also the patch riveted on the central panel between the wheel wells covering the area where the sway braces' cut-outs were usually made. This is due to the fact that the aircraft of this series were initially planned as dive bombers, but were converted into fighters in the factory
Courtesy of Pitești Military Archives via ABC Collection

I.A.R.80-B No. 203, Regimentul 1 Aviație, Popești-Leordeni, August 1950. This aircraft is wearing a single-colour uppersurface camouflage of Olive Green (possibly RLM64) with Light Blue (possibly RLM76) undersurfaces. Pro-Allied markings.

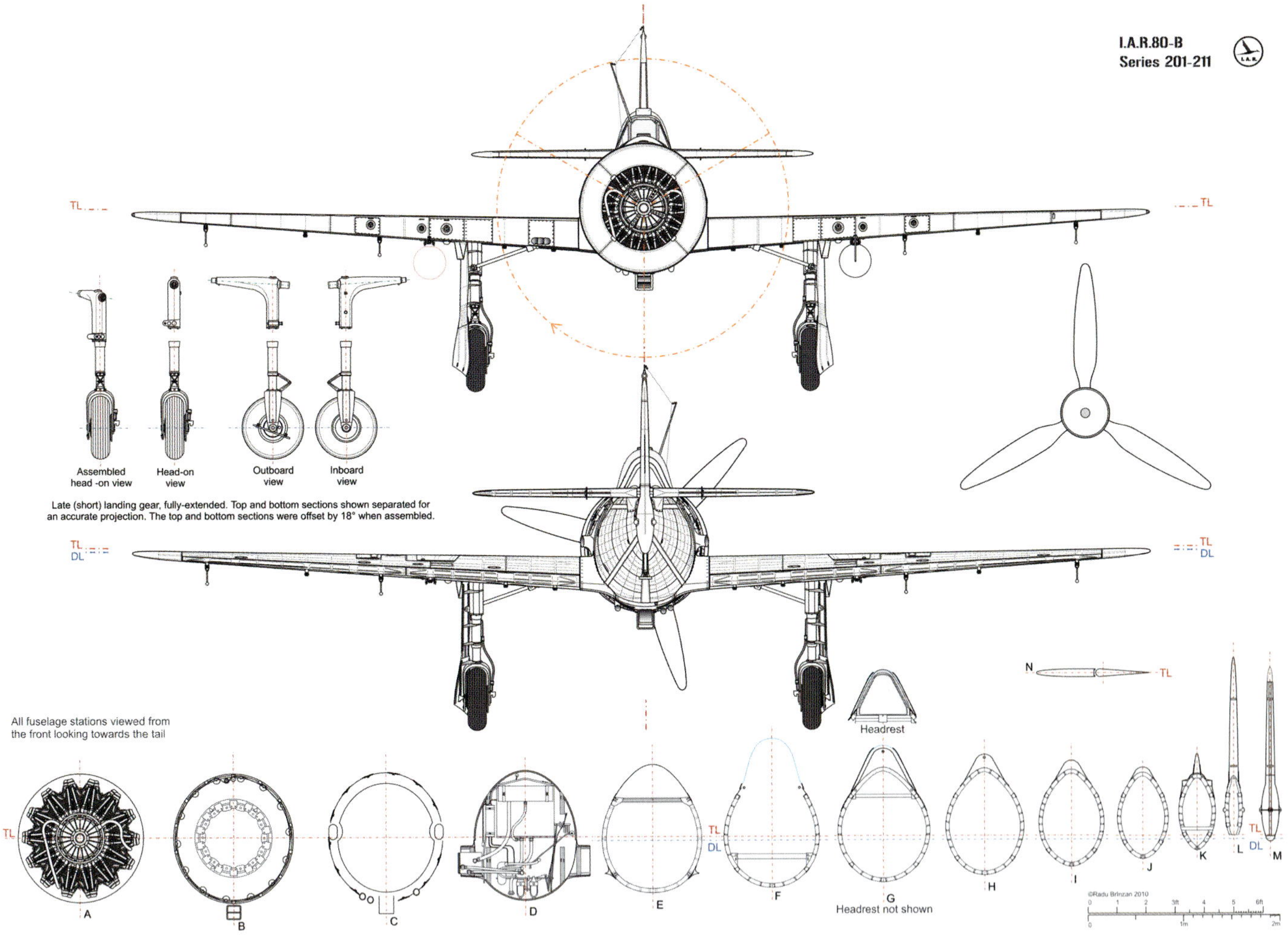

Assembled
head-on view

Head-on
view

Outboard
view

Inboard
view

Late (short) landing gear, fully-extended. Top and bottom sections shown separated for
an accurate projection. The top and bottom sections were offset by 18° when assembled.

Headrest

All fuselage stations viewed from
the front looking towards the tail

Headrest not shown

©Radu Brînzan 2010

A   B   C   D   E   F   G   H   I   J   K   L   M

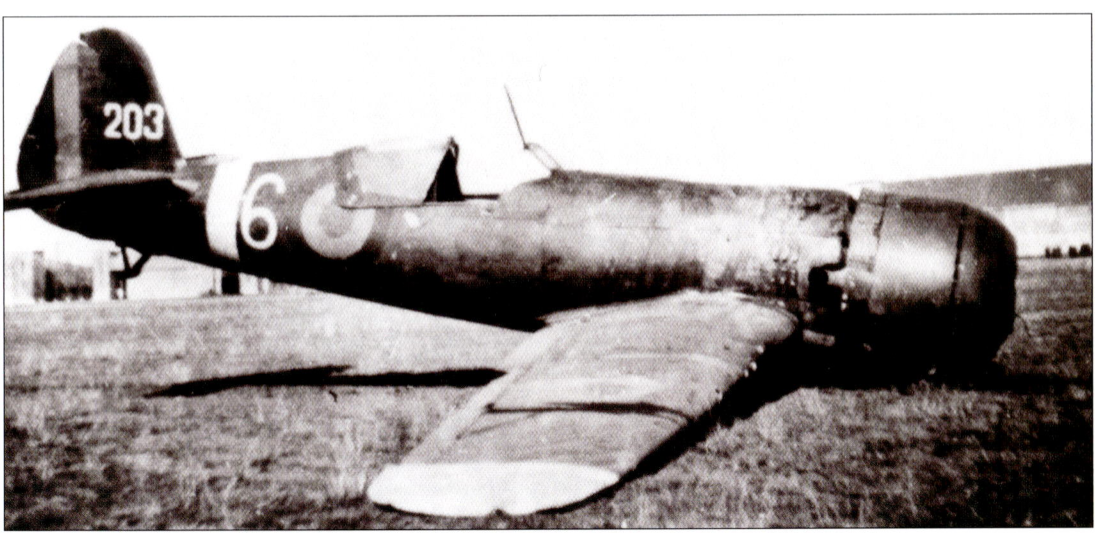

At 8.30 on 24 August 1950, the engine of I.A.R.80-B No. 203 of Reg.1Av. flown by Lt.Maj. Ion Mălăcescu seized during take-off from the Popești-Leordeni airfield. The aircraft crash-landed and was scrapped. This aircraft is still wearing late-war markings. The white '6' is a unit marking and is not standard. Earlier, in the autumn of 1942, when this aircraft was part on Esc.41Vt./Gr.8Vt., it was based on the Tusov airfield and later on the Morozovskaya airfield in the Stalingrad area. On 31 October 1942, this aircraft flown by Slt.Av. Leonid Șotropa crash-landed after it was hit by Soviet AA fire, but was repaired and flew again.
Photo courtesy of ABC collection

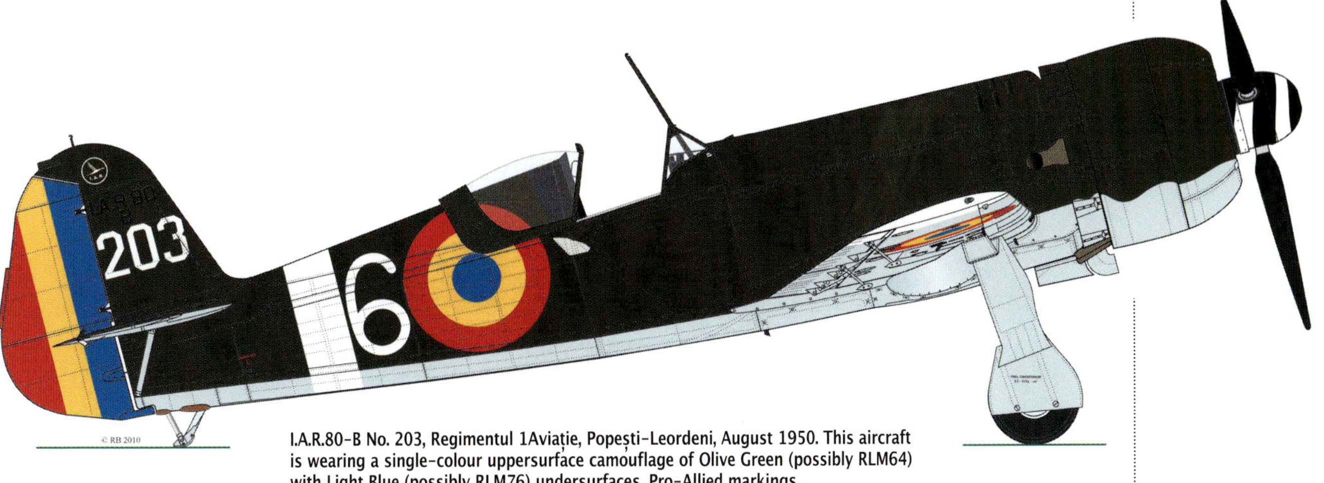

I.A.R.80-B No. 203, Regimentul 1Aviație, Popești-Leordeni, August 1950. This aircraft is wearing a single-colour uppersurface camouflage of Olive Green (possibly RLM64) with Light Blue (possibly RLM76) undersurfaces. Pro-Allied markings.

I.A.R.80-B No. 212 to 230

The aircraft of the 212 to 230 series were initially planned as I.A.R.81-A dive bombers with heavy machine guns, but they were built as fighters. Starting with No.212, the 13.2 mm ammunition payload was increased by 50 rounds to a total of 350. In order to achieve this, some internal and external changes had to be made. The outboard 7.92 mm machine gun was moved slightly further away from the centreline, the ammunition magazine for this gun was moved back and a supplementary ammunition magazine for the 13.2 mm ammunition was placed in this space. The ammunition was brought to the guns through additional feed channels. Externally, the outboard ammunition access panels were enlarged.

Also starting with this series, the wingspan was increased to 11 metres. The structure and skin of the wing were also upgraded. However, it appears that this modification was introduced gradually as many aircraft of this series were fitted with short-span wings, as evidenced by a large number of photographs.

A young pilot poses in the cockpit of I.A.R.80-B No. 214 of Reg.7.Av. Focșani some time after 1950 when the aircraft was painted in the Soviet-style overall grey and wore star-shaped national markings. Note that the pilot wears a German-style white sheepskin fleece jacket this late after the war. The aircraft was retrofitted with an emergency pneumatic canopy opening device as indicated by the circular hatch on the spine. Also, this aircraft was retrofitted with tail plane support struts and was scrapped. Note the short span wing. Courtesy of ABC Collection

I.A.R.80-B No. 220 of Esc.42Vt./Gr.8Vt. was sent to the Morozovskaya airfield near Stalingrad in September 1942. On 29 November 1942, while flown by Adj. Av. Ion Cameniță, this aircraft crash-landed 40 km south of Tazinskaya and was scrapped. Courtesy of ABC Collection

I.A.R.80-B No. 220 of Esc.42Vt./Gr.8Vt. was photographed in the summer 1942 on Tîrgșor airfield before departing for Stalingrad. Note the long span wing. Courtesy of ABC Collection

On 1 August 1943, I.A.R.80-B No. 222 of Esc.62Vt./Gr.6Vt. flown by Lt.Av. Carol Anastasescu scrambled along with the squadron from Pipera airfield to meet the oncoming USAF bombers of Tidal Wave attacking Ploiești. In the battle, Anastasescu shot down one B-24 Liberator and in an attempt to clear the combat area, No. 222 collided in mid-air with another B-24 and both aircraft crashed near the Vega Refinery in Ploiești. Lt.Av. Carol Anastasescu was found alive near the wreckage of his aircraft with a broken back and facial burns and was credited with two victories. This aircraft appears to be fitted with a short-span wing. Courtesy of ABC Collection

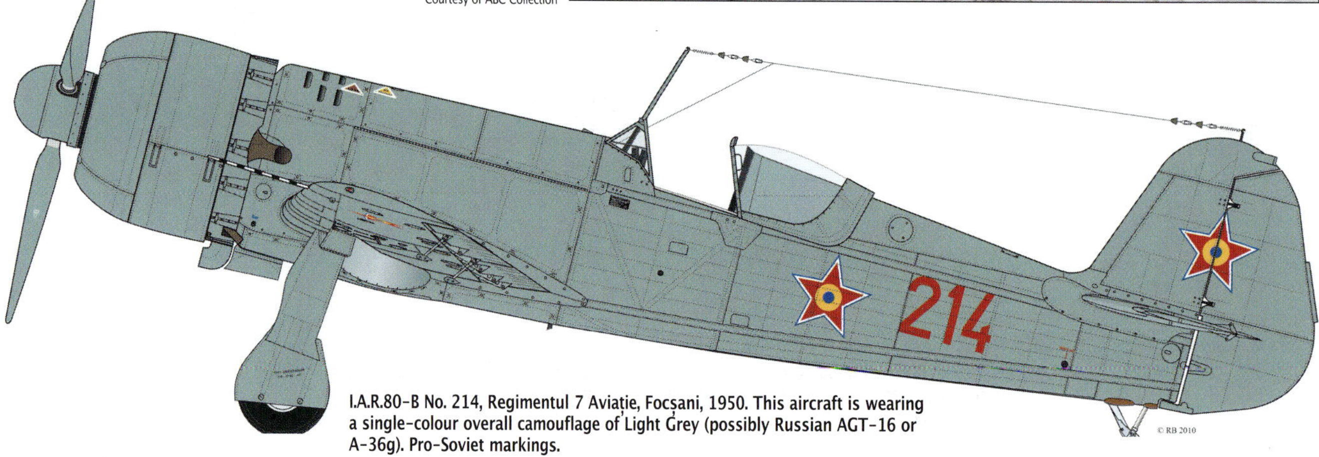

I.A.R.80-B No. 214, Regimentul 7 Aviație, Focșani, 1950. This aircraft is wearing a single-colour overall camouflage of Light Grey (possibly Russian AGT-16 or A-36g). Pro-Soviet markings.

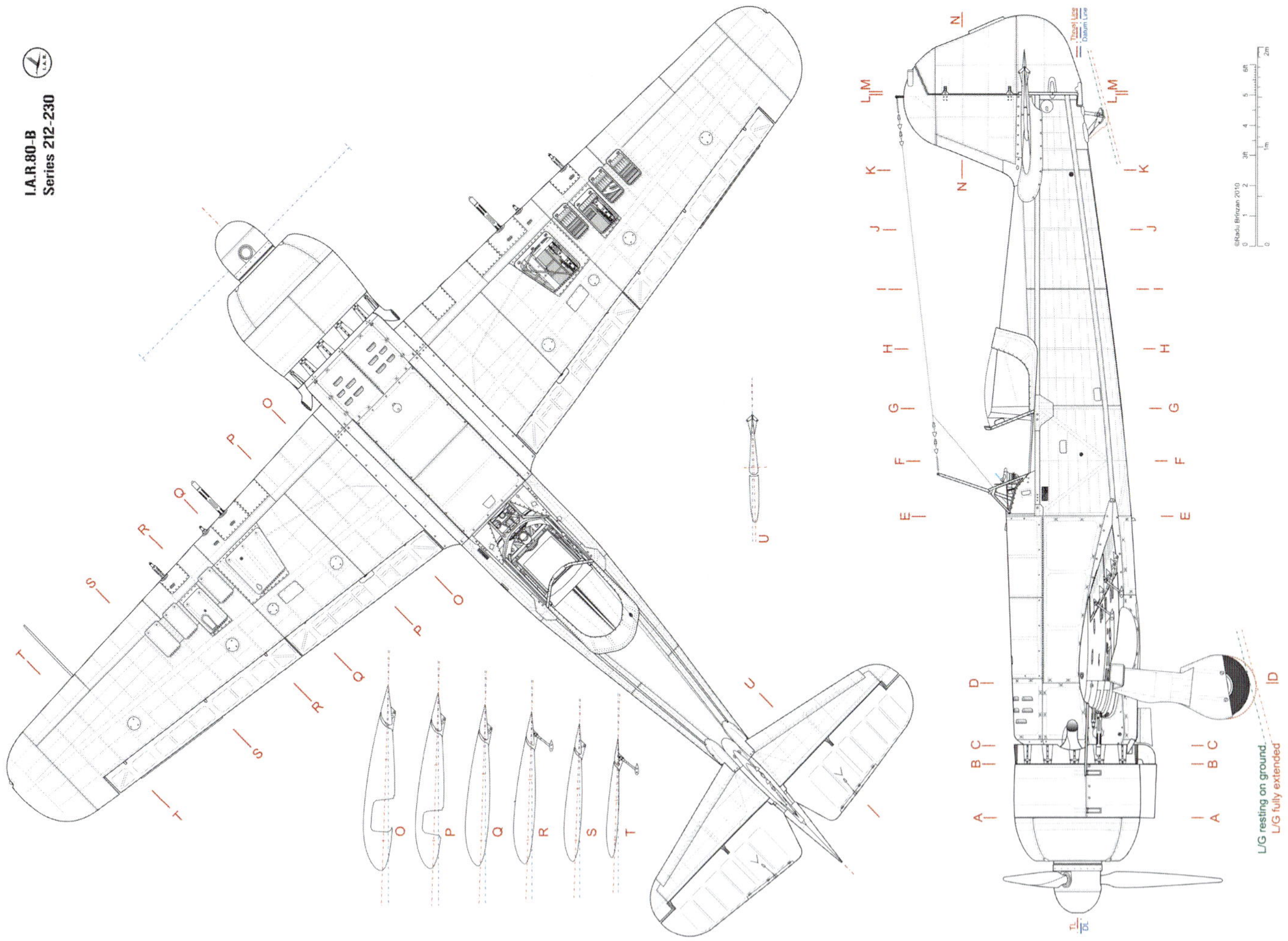

I.A.R.80-B No. 229, 191, 228, 220, 224 and 227 of Esc.42Vt./Gr.8Vt./Flt.2Vt. were photographed in September 1942 on Tîrgşor airfield before departing for Stalingrad. Note the factory-fresh paintwork of these aircraft and compare it to the war-beaten paintwork of No. 220 published elsewhere in this chapter. Courtesy of ABC collection

I.A.R.80-B No. 225 Escadrila 64.Vânătoare, Grupul 1 Vânătoare, Roşiorii de Vede, May 1944. This aircraft is wearing a two-colour wraparound camouflage of Dark Earth and Dark Green with Sky Blue wing and forward fuselage undersurfaces. Eastern Front markings.

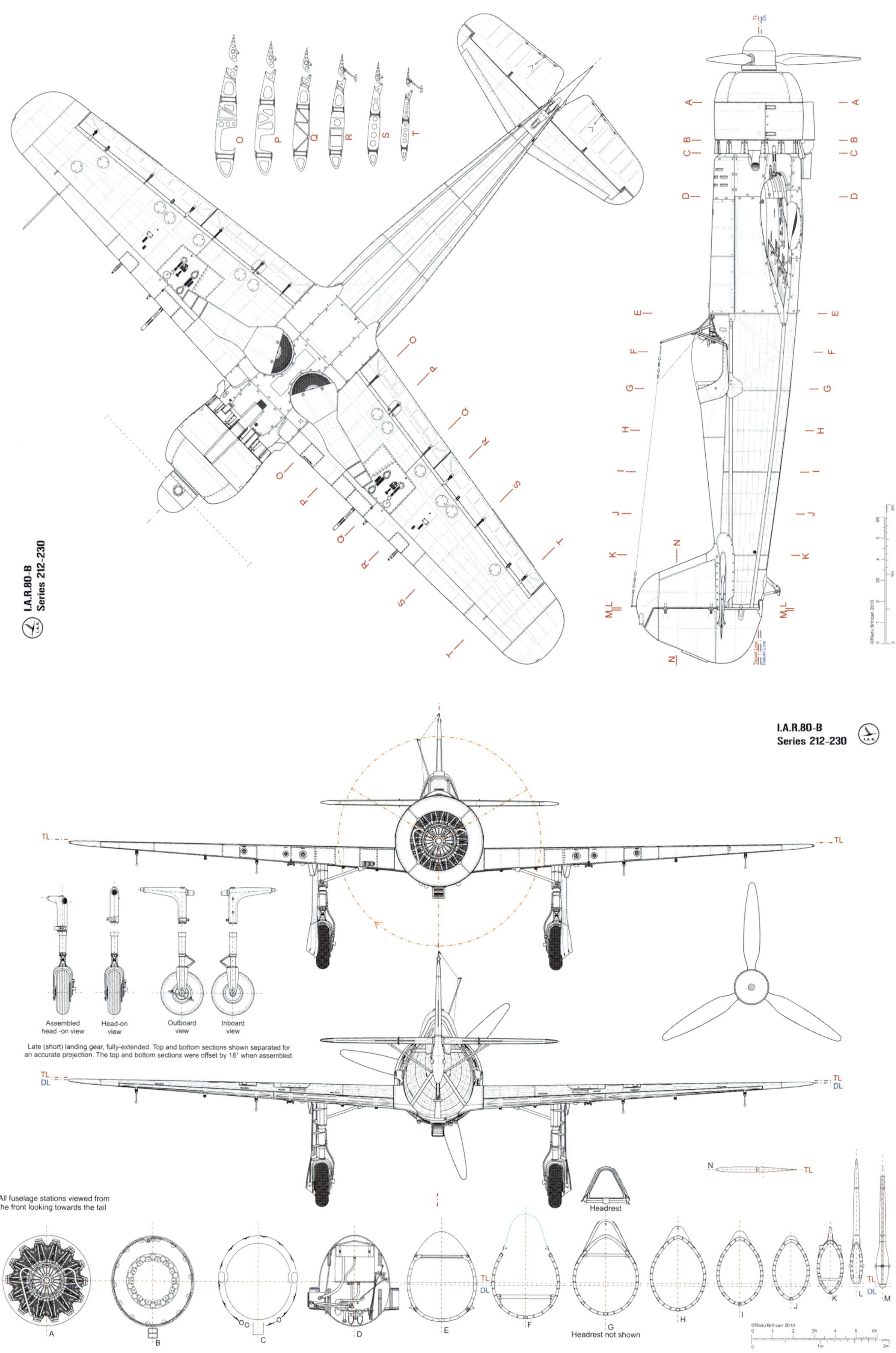

I.A.R.80-B
Series 212-230

I.A.R.80-B
Series 212-230

©Radu Brinzan 2010

Assembled
head-on view

Head-on
view

Outboard
view

Inboard
view

Late (short) landing gear, fully-extended. Top and bottom sections shown separated for
an accurate projection. The top and bottom sections were offset by 18° when assembled.

All fuselage stations viewed from
the front looking towards the tail

Headrest

Headrest not shown

©Radu Brinzan 2010

I.A.R.81 No. 231 to 240

Aircraft of this series had the same armament as that fitted to aircraft of the 151 to 175 with all the airframe modifications introduced up to No.230. The tail planes were fitted with support struts. The flaps were modified to extend to 75° as dive brakes. The second breather tube for the engine gear box introduced with No.201 was removed starting with No.231. This series of aircraft was capable of carrying fuel drop-tanks under the wings.

On 20 January 1944, I.A.R.81 No. 235 of Esc.52Vt. based in Cetatea Albă, crash-landed with Adj. Av. Eugen Jipescu at the controls near the locality of Albăcița, County Tighina in Moldova. The handwritten text on the photograph is 'Direcția de aterizare' [Direction of landing] and "Direcția văii" (Direction of the valley).
Photo courtesy of Pitești Military Archives via ABC Collection

I.A.R.81 No. 235, Escadrila 52 Vânătoare, Cetatea Albă, January 1944. This aircraft is wearing a single-colour uppersurface camouflage of Olive Green (possibly RLM64) with Light Blue (possibly RLM76) undersurfaces. Eastern Front markings.

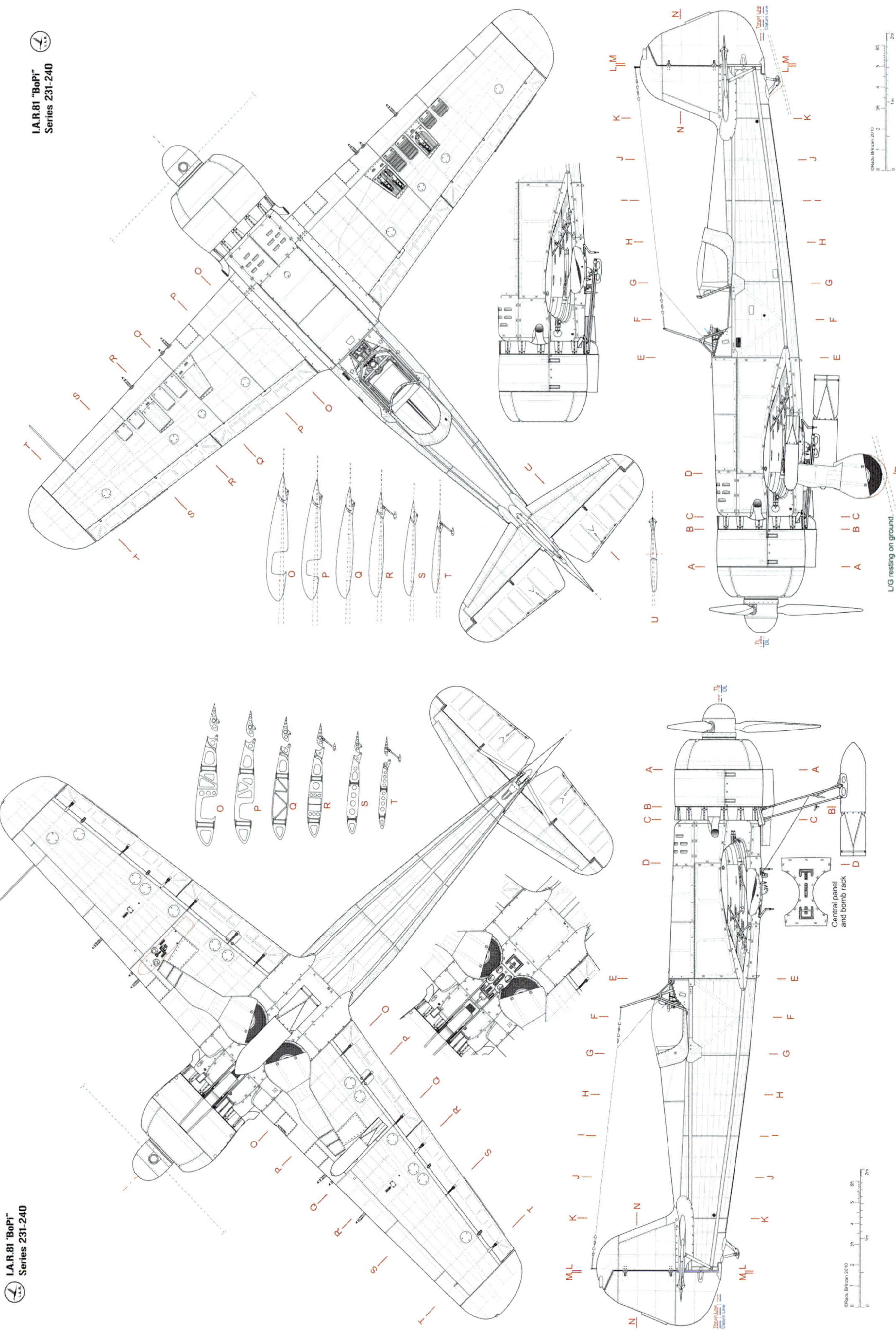

I.A.R.81 "BoPi"
Series 231-240
I.A.R.81 "BoPi"
Series 231-240
L/G resting on ground.
L/G fully extended
Central panel
and bomb rack
©Radu Brinzan 2010
©Radu Brinzan 2010

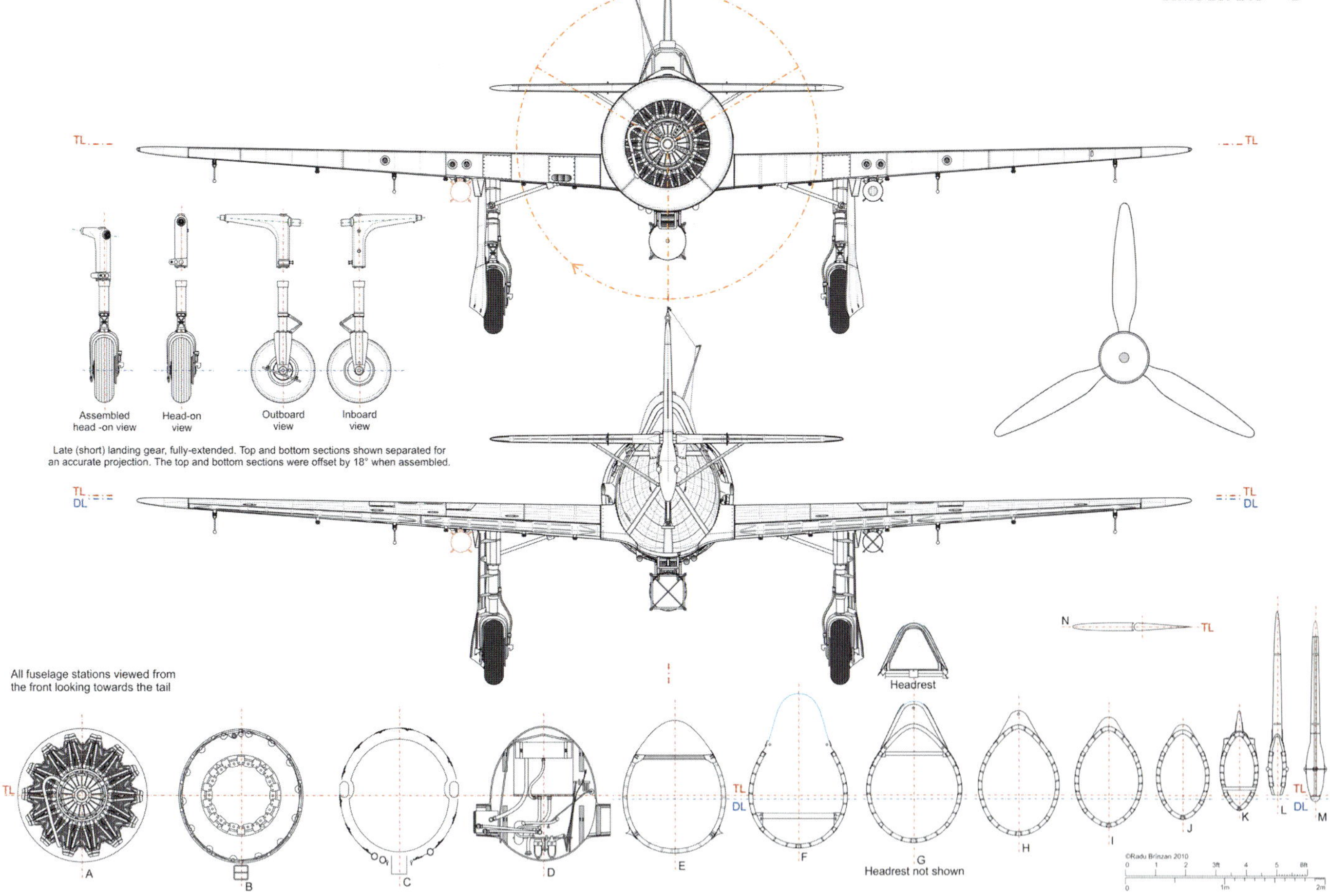

TL — TL

TL
DL

Assembled head-on view

Head-on view

Outboard view

Inboard view

Late (short) landing gear, fully-extended. Top and bottom sections shown separated for an accurate projection. The top and bottom sections were offset by 18° when assembled.

TL DL

All fuselage stations viewed from the front looking towards the tail

Headrest

N — TL

Headrest not shown

A

B

C

D

E

F

G

H

I

J

K

L

M

©Radu Brinzan 2010

I.A.R.81 No. 235 of Esc.52Vt. after the crash-landing on 20 January 1944. This shows the light blue underside of the fuselage and the long spinner. This aircraft was subsequently repaired and flew again. Photo courtesy of Piteşti Military Archives via ABC Collection

I.A.R.80-C No. 241 to 250

Initially planned as a dive bomber armed with MG-FF cannons, the aircraft of this series were manufactured as fighters due to the requirements of the frontline. The armament was modified to 4 x 7.92 mm machine guns with 1,600 rounds and 2 x 20 mm Ikaria (Oerlikon) MG FF cannons, each supplied with 60 drum-fed rounds. This armament made the I.A.R.80-C the I.A.R.80 version with the heaviest punch. Beginning with this series, the intake of the supercharger was fitted with an air filter that could be operated from the cockpit. The tail-plane was fitted with the support struts used on the I.A.R.81 BoPi. The canopy was fitted with a pneumatic ram to assist in opening the canopy while flying at high speed. The wing carriers for the drop tank were modified to carry a maximum payload of 100 kg.

I.A.R.80-C No. 243 of Esc.49Vt./Gr.4Vt. was used on the Eastern Front from Eupatoria airfield in Crimea for the last half of 1943 and the spring of 1944.
Photo courtesy of ABC Collection

A three-quarter frontal view of I.A.R.80-C No. 243 of Esc.49Vt./Gr.4Vt. This photograph was taken with orthochromatic film that made the yellow areas such as the underside of the wingtips and the engine cowl appears very dark.
Photo courtesy of ABC Collection

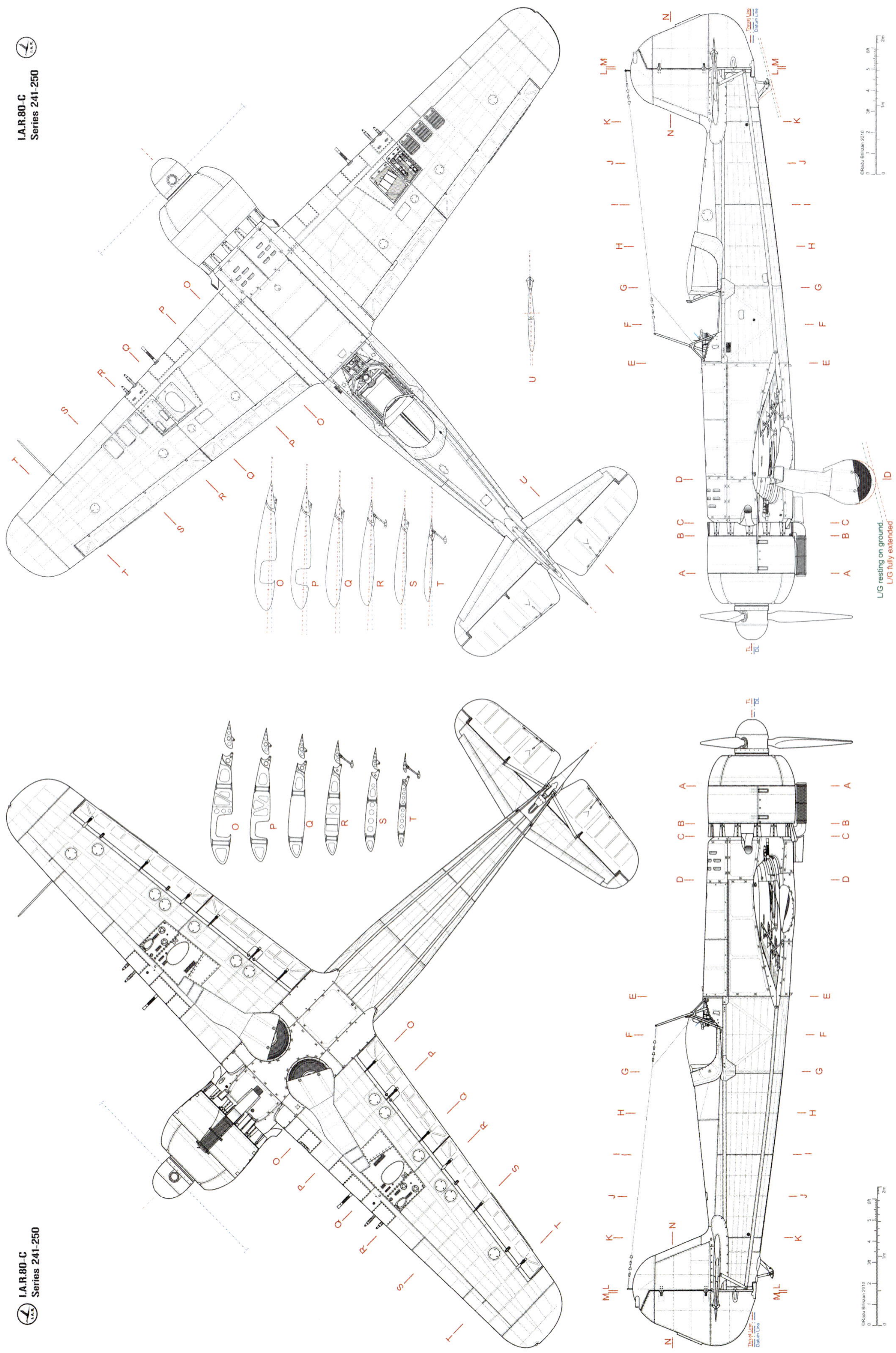

L/G resting on ground.
L/G fully extended

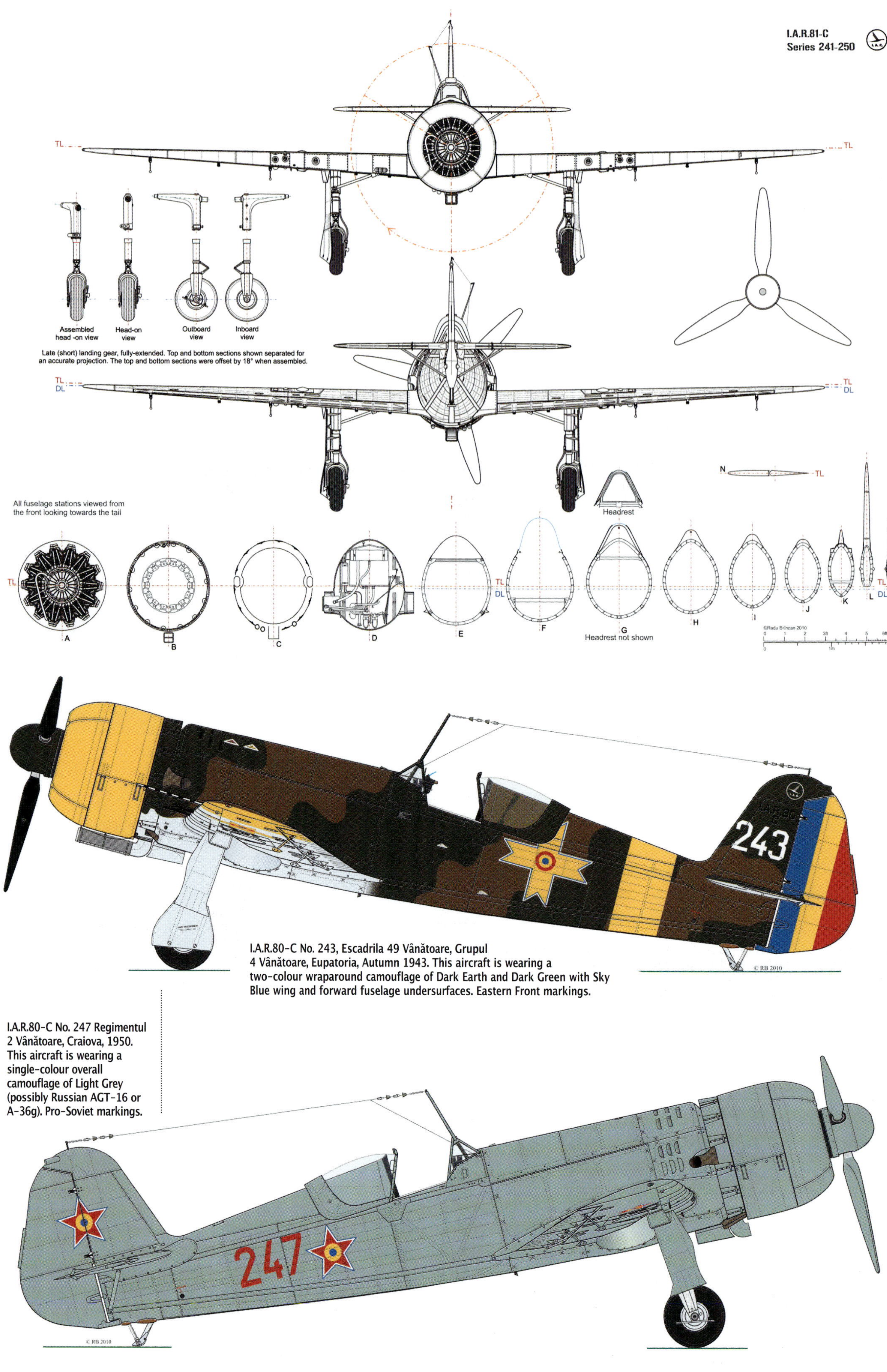

I.A.R.80-C No. 243, Escadrila 49 Vânătoare, Grupul
4 Vânătoare, Eupatoria, Autumn 1943. This aircraft is wearing a
two-colour wraparound camouflage of Dark Earth and Dark Green with Sky
Blue wing and forward fuselage undersurfaces. Eastern Front markings.

I.A.R.80-C No. 247 Regimentul
2 Vânătoare, Craiova, 1950.
This aircraft is wearing a
single-colour overall
camouflage of Light Grey
(possibly Russian AGT-16 or
A-36g). Pro-Soviet markings.

I.A.R.80-C No. 251 to 290

Initially planned as a dive bomber armed with MG-FF cannons, the aircraft of this series were manufactured as fighters due to the requirements of the frontline. Aircraft of this series had the same armament as that fitted to aircraft of the 241-250 series with all the airframe modifications introduced up to No.250. The shape of the pilot's headrest cushion was changed. The internal fuel tanks were coated with a self-sealing material. A second oil cooler was added to the left wing, symmetrical to the existing oil cooler in the right wing. Extra cooling gills of a unique type used only on this series were added on the fuselage sides behind the exhausts to assist oil cooling.

I.A.R.80-C No. 259 of Gr.6Vt./Flt.2Vt. was photographed on Someșeni airfield in the autumn of 1944. The nose of a wingless I.A.R. 27 trainer is visible at the right edge of the photo. Note the late-war markings and the long spinner.
Courtesy of ABC Collection

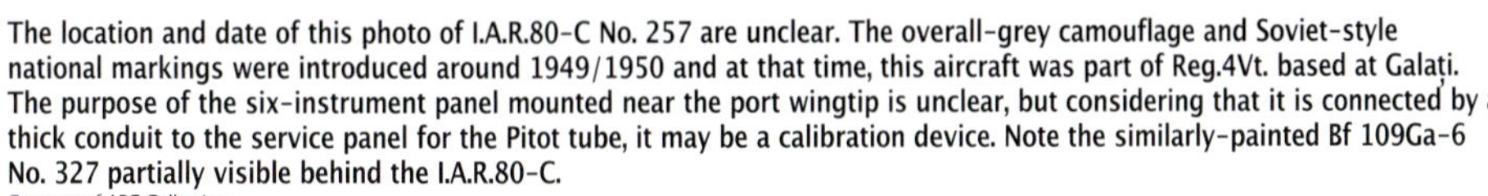

The location and date of this photo of I.A.R.80-C No. 257 are unclear. The overall-grey camouflage and Soviet-style national markings were introduced around 1949/1950 and at that time, this aircraft was part of Reg.4Vt. based at Galați. The purpose of the six-instrument panel mounted near the port wingtip is unclear, but considering that it is connected by a thick conduit to the service panel for the Pitot tube, it may be a calibration device. Note the similarly-painted Bf 109Ga-6 No. 327 partially visible behind the I.A.R.80-C.
Courtesy of ABC Collection

A closer view of the nose of I.A.R.80-C No. 279. Note the unusual cooling gills on the nose sides, typical for this series of I.A.R.80. On 28 August 1943, this aircraft was sent to I.A.R. Brașov for overhaul and in March 1944 it was transferred to Esc.49Vt. based in Saki in Crimea. On 9 April 1944, this aircraft lost its propeller in flight over the sea and glided to Eupatoria airfield. By the end of the month, the Romanian armed forces were on retreat from the Eastern Front and the aircraft was scheduled to be transported by ship back to Romania, but it was abandoned on Saki airfield.
Courtesy of ABC Collection

I.A.R.80-C No. 257, Regimentul 4 Vânătoare, Galați, 1950. This aircraft is wearing a single-colour overall camouflage of Light Grey (possibly Russian AGT-16 or A-36g). Pro-Soviet markings

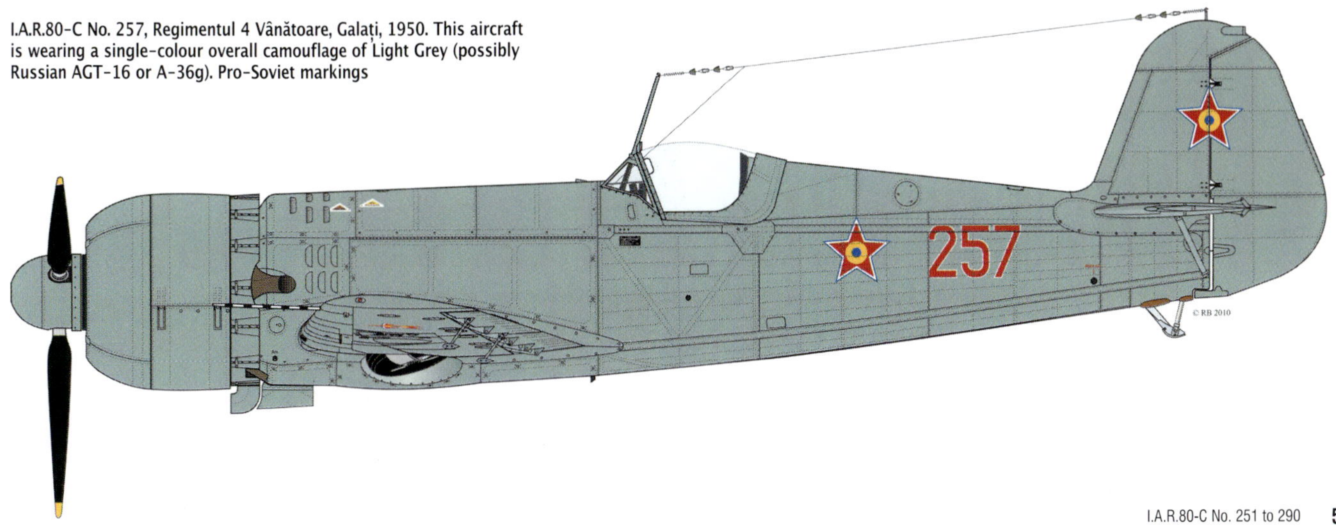

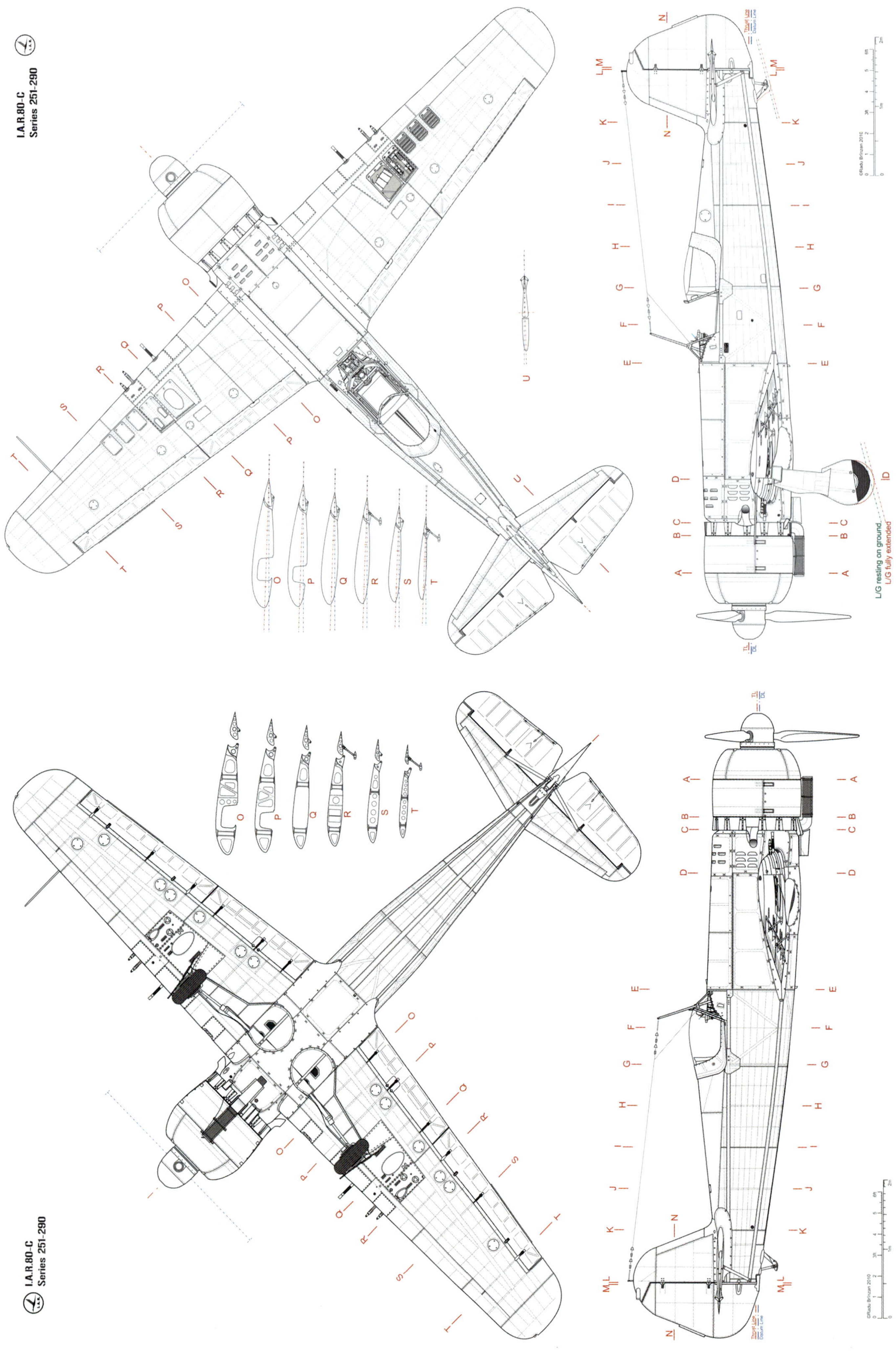

I.A.R.80-C
Series 251-290
N
L M
K
J
I
H
F G
E
D
A B C
L/G resting on ground
L/G fully extended
©Radu Brînzan 2010

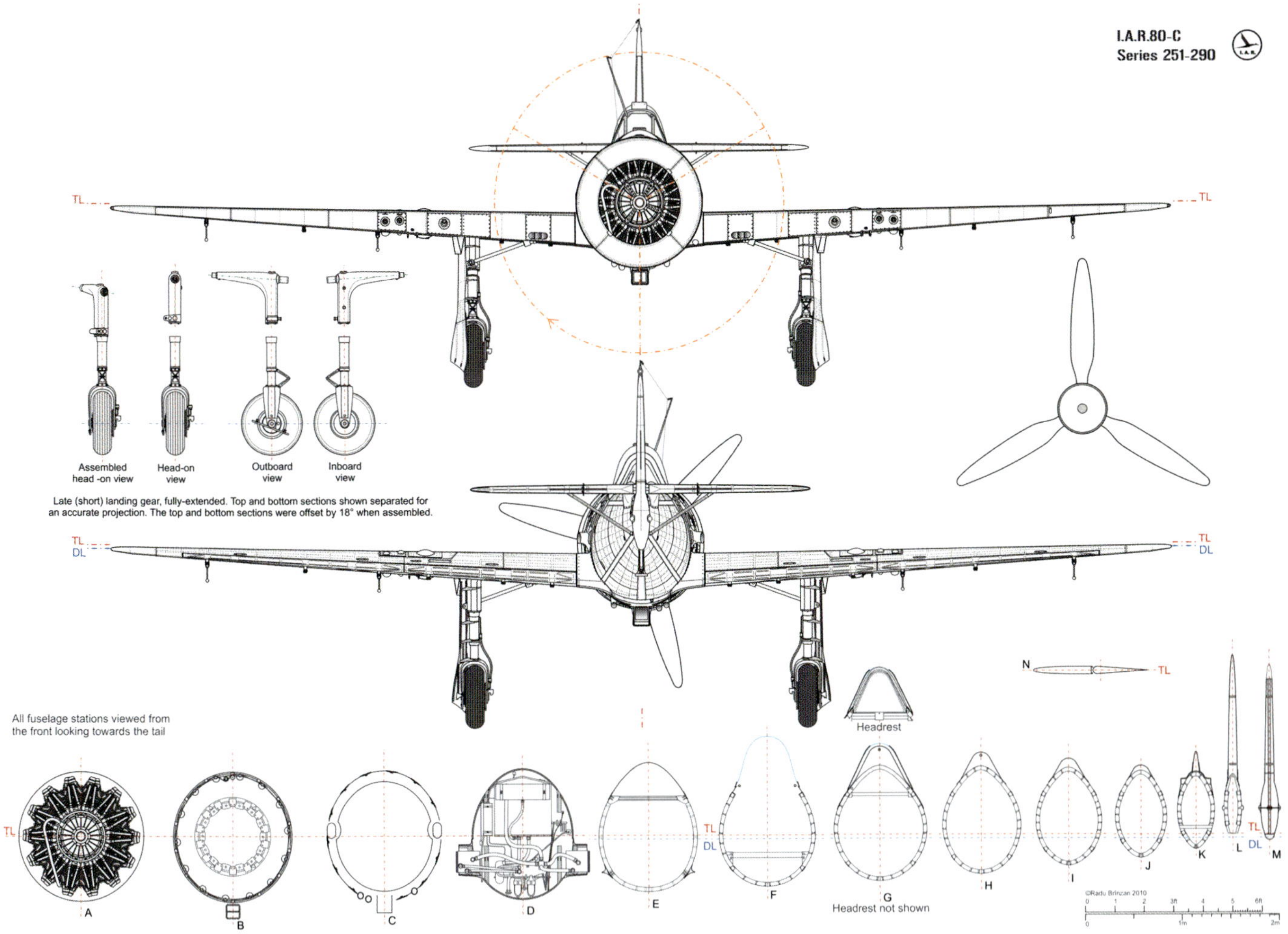

Assembled
head -on view

Head-on
view

Outboard
view

Inboard
view

Late (short) landing gear, fully-extended. Top and bottom sections shown separated for
an accurate projection. The top and bottom sections were offset by 18° when assembled.

All fuselage stations viewed from
the front looking towards the tail

Headrest

N

Headrest not shown

©Radu Brînzan 2010

A   B   C   D   E   F   G   H   I   J   K   L   M

I.A.R.80-C No. 270 was part of Esc.46Vt./Gr.4Vt. based on Cetatea Albă airfield. Note the
very distinctive cooling gills on the fuselage behind the exhausts. This type of cooling
gills was used only on the aircraft of this series.
Courtesy of ABC Collection

I.A.R.80-B No. 279 of Esc.45Vt. was flown by Lt.Av.
Ion Bârlădeanu from Târgșor airfield in August 1943.
This aircraft was personalised with a cartoon of
'Tenebras, King of Cons', visible on the starboard side
of the nose.
Courtesy of ABC Collection

I.A.R.80-B No. 279, Escadrila 45 Vânătoare, Târgșor, August 1943. This aircraft is wearing a single-colour
uppersurface camouflage of Olive Green (possibly RLM64) with Light Blue (possibly RLM76) undersurfaces.
Eastern Front markings. The cartoon of "Tenebras, Regele Coțcarilor" [Tenebras King of Cons] drawn on the
nose of I.A.R.80-C No. 279 was reportedly based on a popular wartime comic book character.

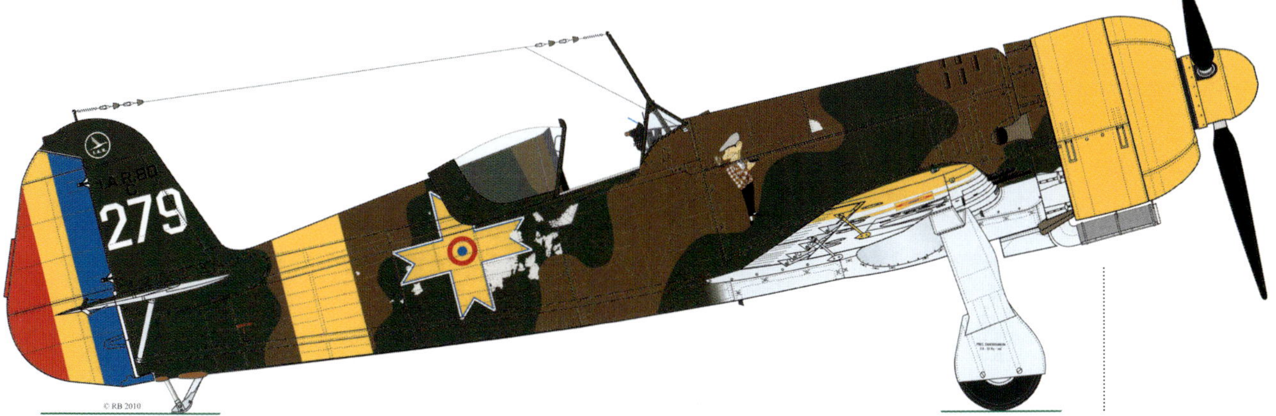

I.A.R.80-B No. 291 to 300

The aircraft of the 291 to 300 series were initially planned as I.A.R.81-A dive bombers with heavy machine guns, but they were built as fighters with the same armament as that fitted to aircraft of the 212-230 series with all the airframe modifications introduced up to No.290.

A pilot poses in front of an unidentified I.A.R.80 of the 291-300 series as indicated by the blisters for the heavy machine gun visible under the wing and the second oil radiator. This aircraft is wearing the Soviet-style overall grey camouflage scheme and star-type national insignia.
Photo courtesy of ABC collection

I.A.R.80-B No. 295 after an accident in June 1947. The hourglass-shaped object behind the wing is the central fuselage panel between the wheel wells. Between the winter of 1943 and the autumn of 1944, this aircraft was part of Esc.64Vt./Gr.1Vt. based in Roșiorii de Vede. On 21 April 1944, it was hit during a battle with USAF aircraft. It was repaired by ASAM Cotroceni and returned to the same unit in late-August. On 1 September 1944, it was hit during a battle with Luftwaffe aircraft. After it was repaired by ASAM Cotroceni, it became an unarmed school aircraft in 1945.
Photo courtesy of Pitești Military Archives via ABC Collection

I.A.R.80-B No. 295, Flotila 3 Asalt, Brașov, June 1947. This aircraft is wearing a single-colour uppersurface camouflage of Olive Green (possibly RLM64) with Light Blue (possibly RLM76) undersurfaces.
Pro-Allied markings

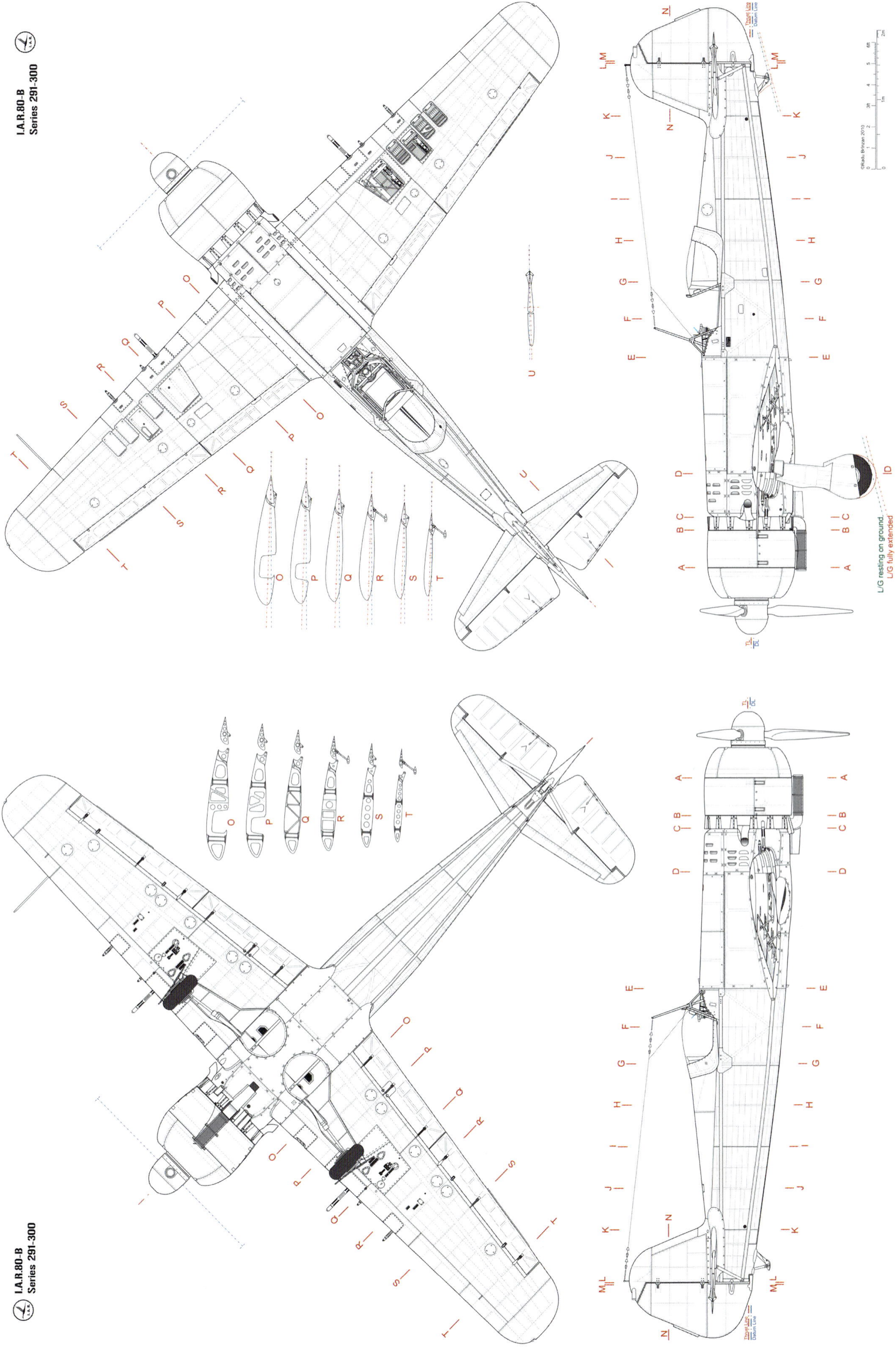

I.A.R.80-B
Series 291-300
L/G resting on ground
L/G fully extended
©Radu Brinzan 2010

I.A.R.80-B
Series 291-300

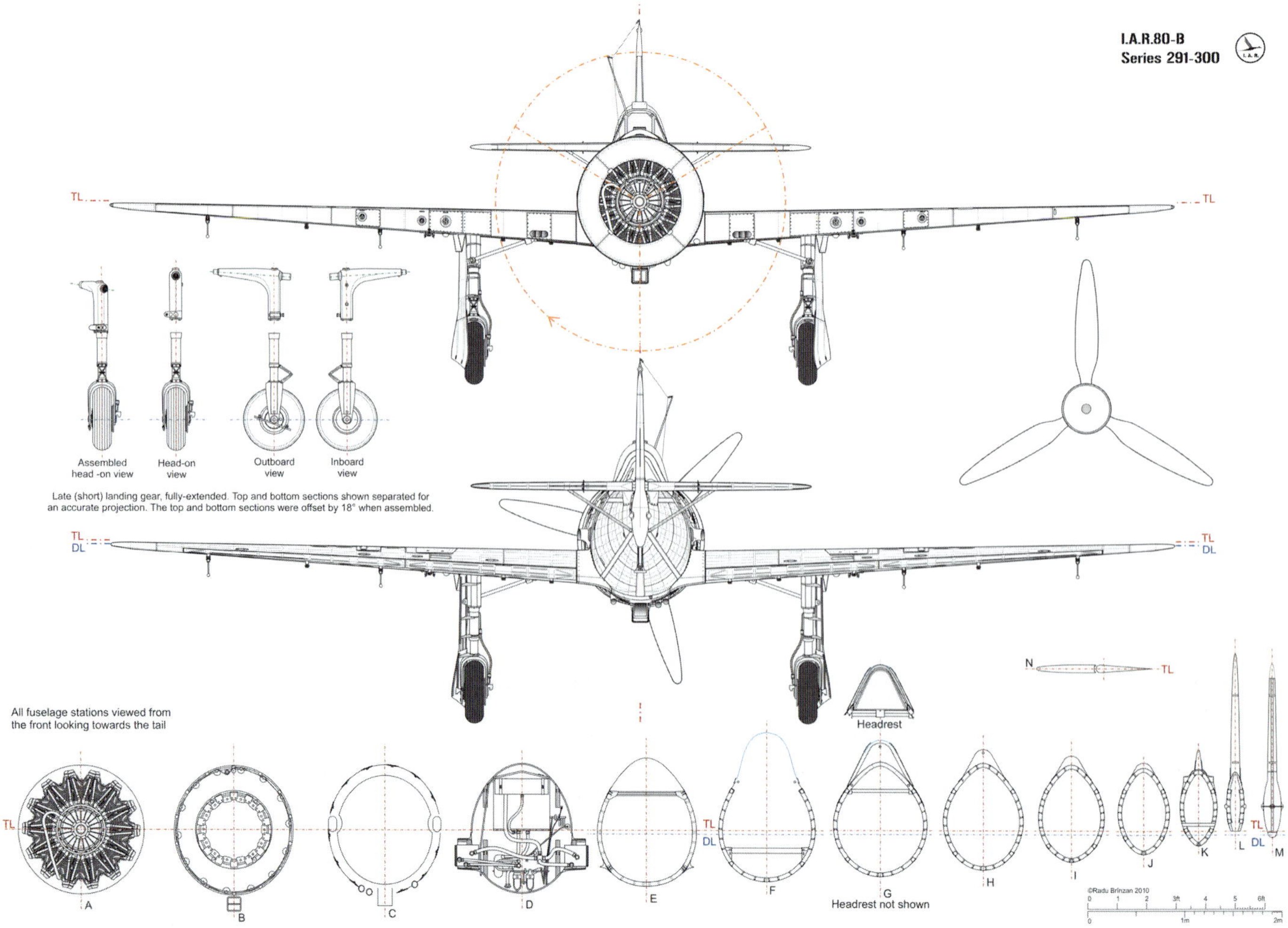
TL
TL
Assembled head -on view
Head-on view
Outboard view
Inboard view
Late (short) landing gear, fully-extended. Top and bottom sections shown separated for an accurate projection. The top and bottom sections were offset by 18° when assembled.
TL
DL
TL
DL
N
TL
Headrest
All fuselage stations viewed from the front looking towards the tail
TL
A
B
C
D
E
TL
DL
F
G
Headrest not shown
H
I
J
K
L
TL
DL
M
©Radu Brînzan 2010

On 24 October 1947, I.A.R.80-B No. 291 of Şc.Mil.Av. Turda with Slt.Av. Remus Munteanu at the controls was damaged during landing. The aircraft was scrapped in March 1948 in Clinceni.
Photo courtesy of Piteşti Military Archives via ABC Collection

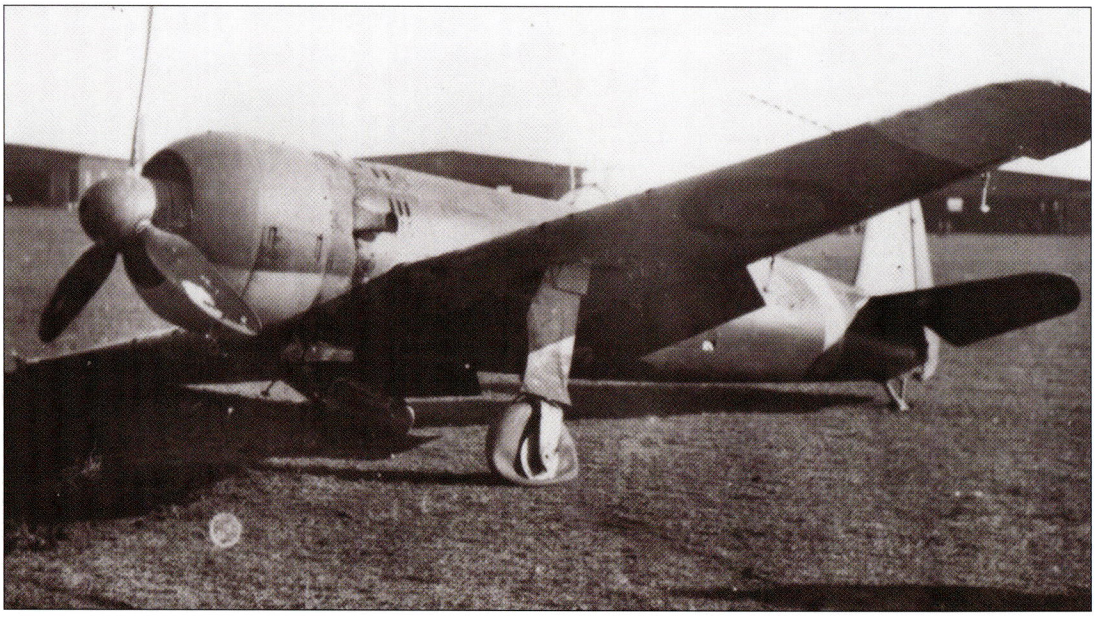

I.A.R.80-B No. 291, Şcoala Militară de Aviaţie, Turda, October 1947. This aircraft is wearing a single-colour uppersurface camouflage of Olive Green (possibly RLM64) with Light Blue (possibly RLM76) undersurfaces.
Pro-Allied markings

291
© RB 2010

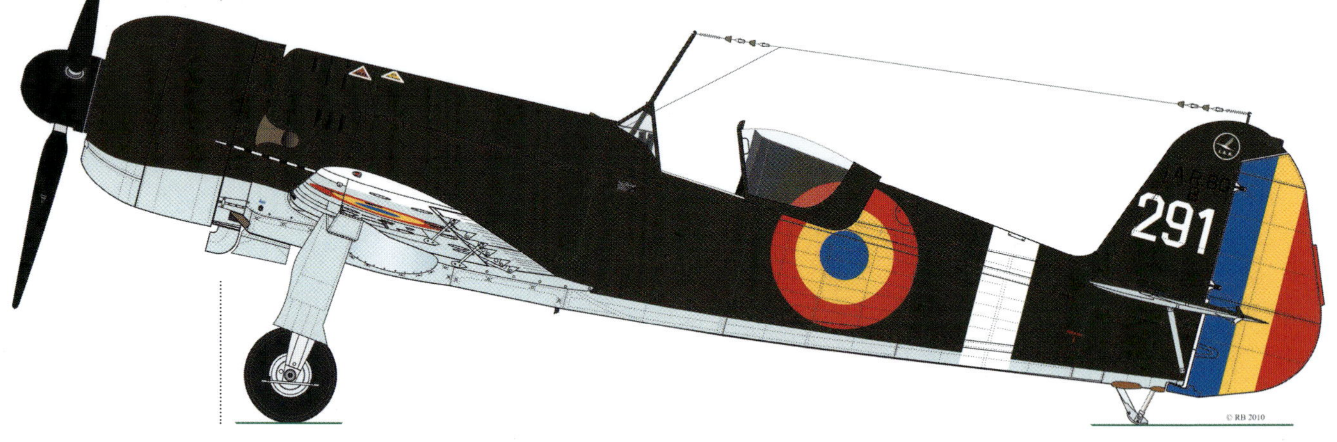

I.A.R.81-C No. 301 to 450

Derived from the 291-300 series, this was the definitive version of the aircraft and was built in the largest number accounting for one third of the entire production run of I.A.R.80 and I.A.R.81. As the name indicates, this was designed from the very beginning as a dive bomber and there is photographic evidence that many (if not all) aircraft of this series left the factory equipped with bomb-carrying equipment. However, the central bomb carrying equipment was removed from the majority of aircraft, which were subsequently used as pure fighters.

The armament was changed to 2 x 20 mm MG151/20 Mauser cannons with 350 belt-fed rounds, and 2 x 7.92 mm FN Browning machine guns with 1,400 rounds. The armament was

I.A.R.81-C No. 320 of Esc.61Vt./Gr.6Vt. was photographed on a snow-covered Popești-Leordeni airfield in the early spring of 1944. Note the large white triangle marking on the fuselage and the landing gear doors missing from the left leg Photo courtesy of ABC Collection

charged pneumatically and fired electrically. The front armour of the windscreen was increased to 75 mm. The location of the wing-mounted bomb and drop tank carriers was changed to a position further outboard. A modified air filter was added to the supercharger intake with the intake shutter linked to the landing gear, which closed it automatically when the landing gear extended.

Lt.Av. Mircea Dumitrescu the Comandant of Esc.61Vt./Gr.6Vt. wearing a German sheepskin flight jacket was photographed looking at the 'Bambi' emblem of Esc.61Vt./Gr.6Vt. painted on I.A.R.81-C No. 320.
Photo courtesy of ABC Collection

Lt.Av. Mircea Dumitrescu, is boarding I.A.R.81-C No. 320 of Esc.61Vt./Gr.6Vt. on Popești-Leordeni airfield during the summer of 1944. The large white triangle was a formation marking and indicated that this was the aircraft flown by the leader of the battle formation. Note the repainted rivet lines on the rear fuselage
Photo courtesy of Jose Fernandez

At 16.55 on 24 September 1943 Lt.Av. Constantin Tulică, the young Sublocotenent smiling in this photo with his left hand in his pocket, took off on a scout flight in I.A.R.81-C No. 318 of Esc.59Vt./Gr.6Vt. from Tîrgșor airfield and force-landed on an embankment road near the locality of Golești (Muscel) due to a seized engine. The right landing gear leg and the propeller were damaged. Note the spiral on the spinner and the open right-wing cannon bay
Photo courtesy of Pitești Military Archives via ABC Collection

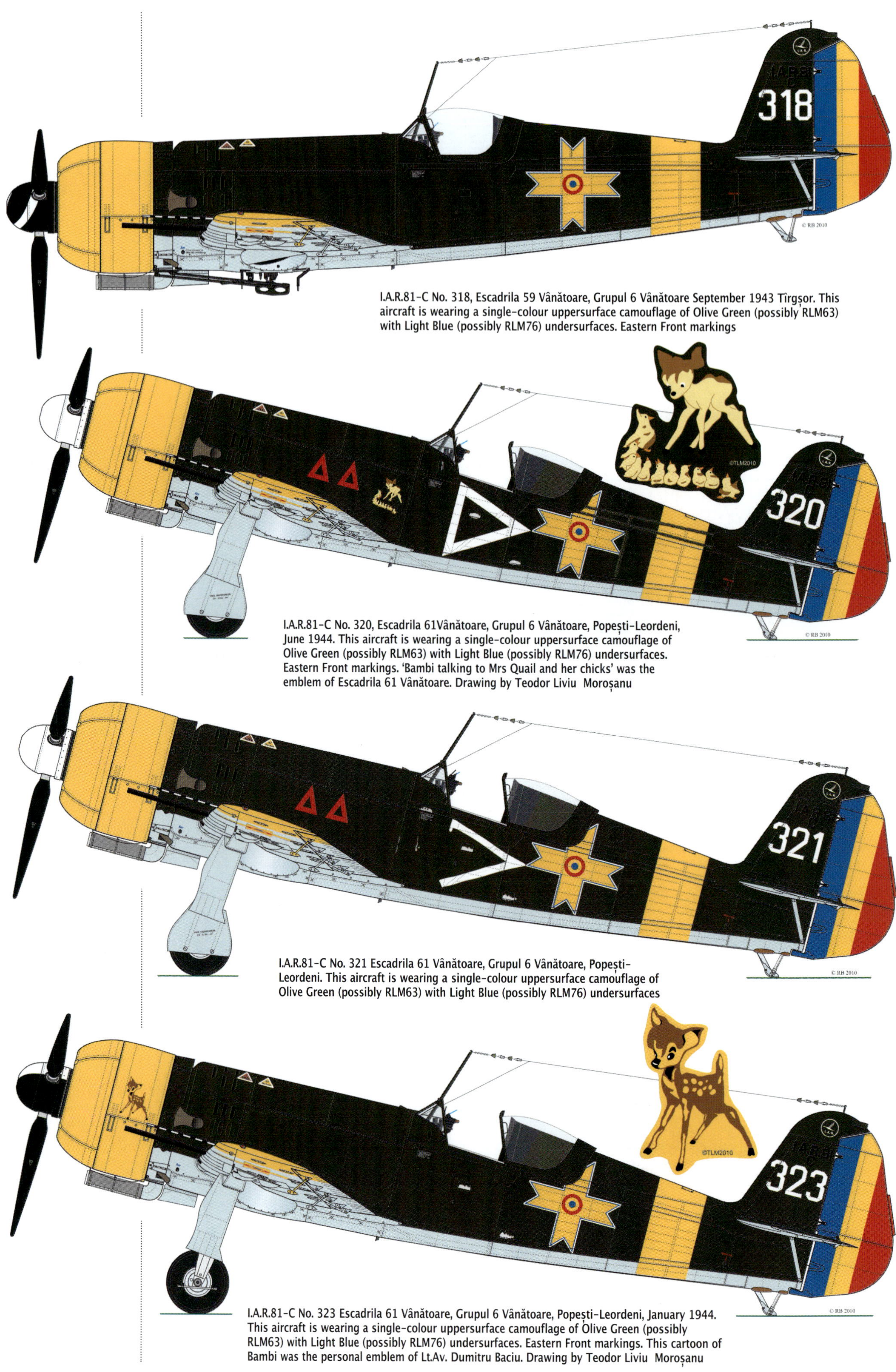

I.A.R.81-C No. 318, Escadrila 59 Vânătoare, Grupul 6 Vânătoare September 1943 Tîrgșor. This aircraft is wearing a single-colour uppersurface camouflage of Olive Green (possibly RLM63) with Light Blue (possibly RLM76) undersurfaces. Eastern Front markings

I.A.R.81-C No. 320, Escadrila 61Vânătoare, Grupul 6 Vânătoare, Popești-Leordeni, June 1944. This aircraft is wearing a single-colour uppersurface camouflage of Olive Green (possibly RLM63) with Light Blue (possibly RLM76) undersurfaces. Eastern Front markings. 'Bambi talking to Mrs Quail and her chicks' was the emblem of Escadrila 61 Vânătoare. Drawing by Teodor Liviu Moroșanu

I.A.R.81-C No. 321 Escadrila 61 Vânătoare, Grupul 6 Vânătoare, Popești-Leordeni. This aircraft is wearing a single-colour uppersurface camouflage of Olive Green (possibly RLM63) with Light Blue (possibly RLM76) undersurfaces

I.A.R.81-C No. 323 Escadrila 61 Vânătoare, Grupul 6 Vânătoare, Popești-Leordeni, January 1944. This aircraft is wearing a single-colour uppersurface camouflage of Olive Green (possibly RLM63) with Light Blue (possibly RLM76) undersurfaces. Eastern Front markings. This cartoon of Bambi was the personal emblem of Lt.Av. Dumitru Baciu. Drawing by Teodor Liviu Moroșanu

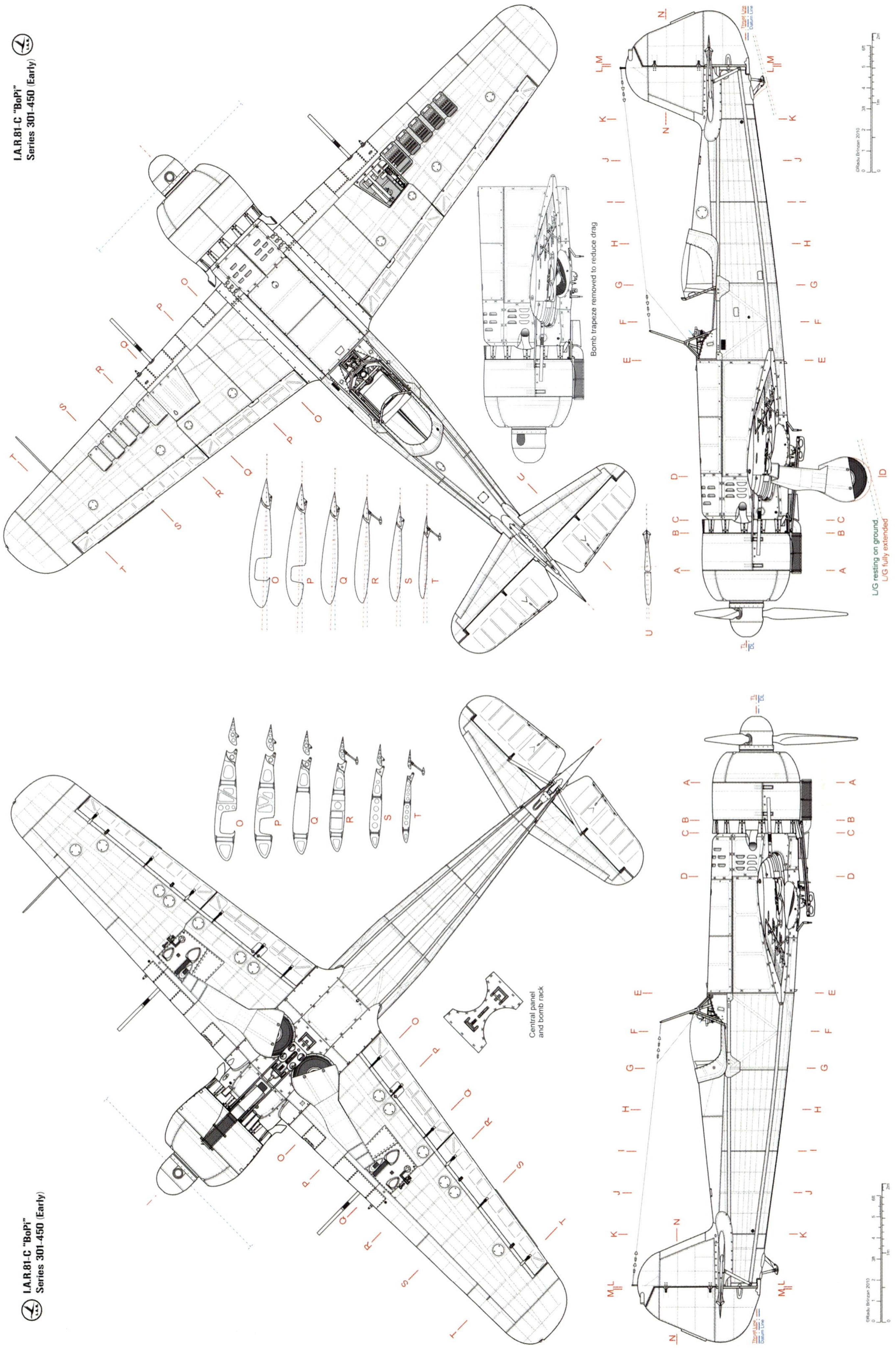

I.A.R.81-C "BoPi"
Series 301-450 (Early)
I.A.R.81-C "BoPi"
Series 301-450 (Early)
Bomb trapeze removed to reduce drag
Central panel
and bomb rack
L/G resting on ground
L/G fully extended
©Radu Brinzan 2010
©Radu Brinzan 2010

Lt.Av. Dumitru Baciu of Esc.61Vt./Gr.6Vt. showing 'Bambi', his personal emblem painted on the cowl of I.A.R.81-C No. 323 in January 1944. Later in 1944, Lt.Av. Baciu was credited with shooting down one P-38 on 10 June and one B-17 on 23 June while flying another I.A.R.81-C.
Photo courtesy of ABC Collection

A number of pilots consult a map in front of I.A.R.81-C No. 329 of Esc.61.Vt./Gr.6Vt. The pilot wearing the white overalls is Lt.Av. Mircea Dumitrescu, the Comandant of Esc.61Vt. On 31 May 1944, I.A.R.81-C No. 329 flown by Adj.Av. Victor Petric was hit in the left wing and flap during a battle with USAF aircraft but managed to return to the airfield.
Photo courtesy of Jose Fernandez

At 16.32 on 16 October 1947, I.A.R.81-C No. 345 of Şcoala Miltară Aviaţie flown by Slt.Av. Anatolie Grunju was damaged during landing on Turda airfield. The previous yellow band is showing through the worn-out white band and repainted green surrounding areas. The two red triangles still faintly visible on the nose are the vestiges of erstwhile Esc.61/Gr.6Vt. formation markings. Interestingly, although the aircraft is not armed, it still has a gunsight.
Photo courtesy of Piteşti Military Archives via ABC Collection

Lt.Av. Teodor Nicolaescu stands in the cockpit of I.A.R.81-C No. 321 Esc.61Vt./Gr.6Vt. in Popeşti-Leordeni. During the morning of 31 May 1944, I.A.R.81-C No. 321 flown by Lt.Av. Teodor Nicolaescu engaged in an aerial battle with four USAF P-51 Mustangs. After expending his ammunition, in an attempt to escape, Lt.Av. Nicolaescu entered a steep dive from 6,000 metres. During the dive, the propeller flew off and he had to glide the powerless aircraft to a belly-landing near the Bucharest bypass road.
Photo courtesy of ABC Collection

Lt.Av. Dumitru Baciu in a flight suit is briefed by an officer in a trench coat in front of I.A.R.81-C No. 323 of Esc.61Vt./Gr.6Vt. on Popeşti-Leordeni airfield in January 1944. The bottom doors of the landing gear were removed to prevent the wheels from becoming clogged with snow and mud during take-off and landing.
Photo courtesy of ABC Collection

I.A.R.81-C "BoPi"
Series 301-450 (Early)

Assembled head-on view

Head-on view

Outboard view

Inboard view

Late (short) landing gear, fully-extended. Top and bottom sections shown separated for an accurate projection. The top and bottom sections were offset by 18° when assembled.

All fuselage stations viewed from the front looking towards the tail

Headrest

Headrest not shown

©Radu Brinzan 2010

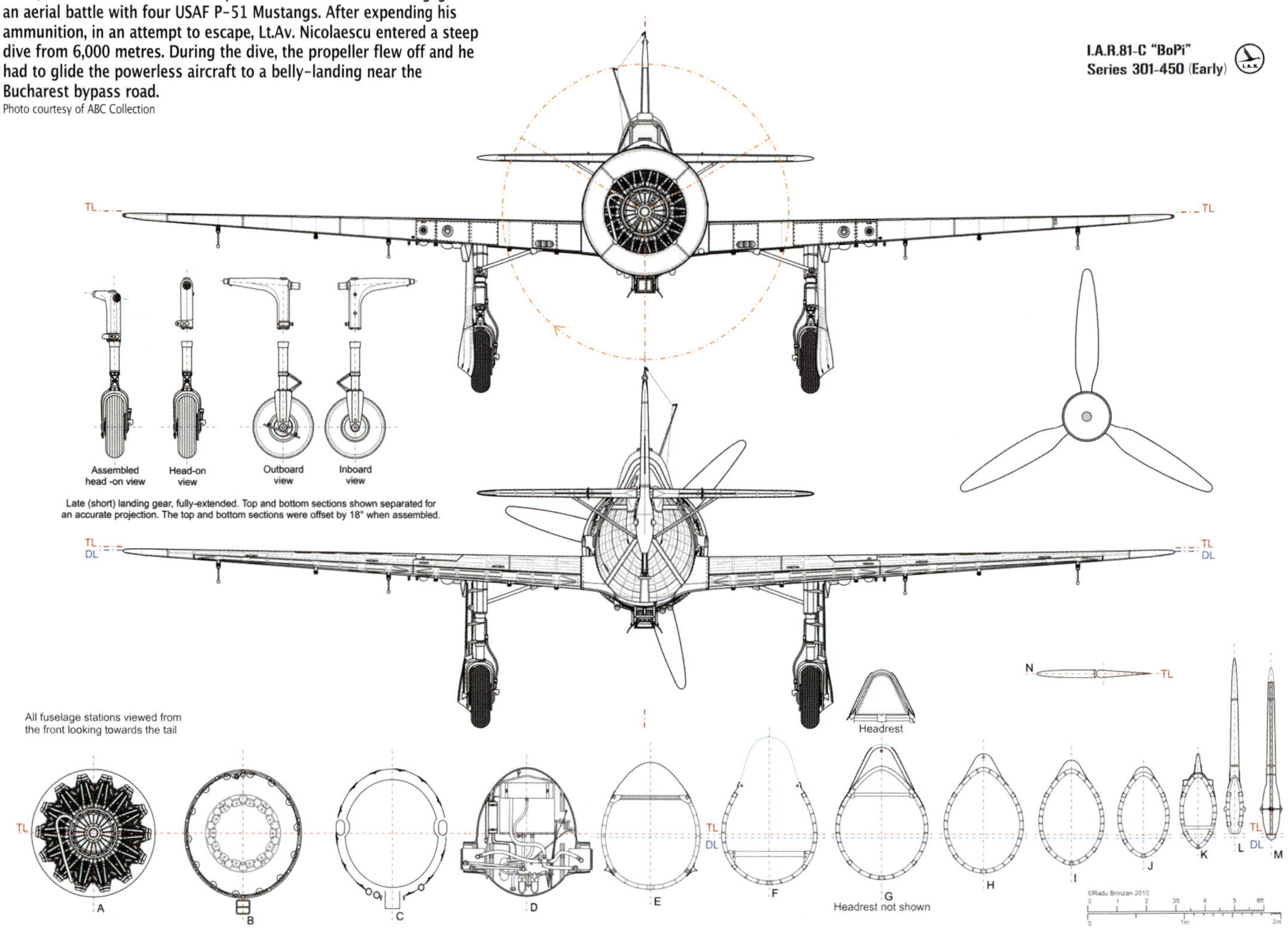

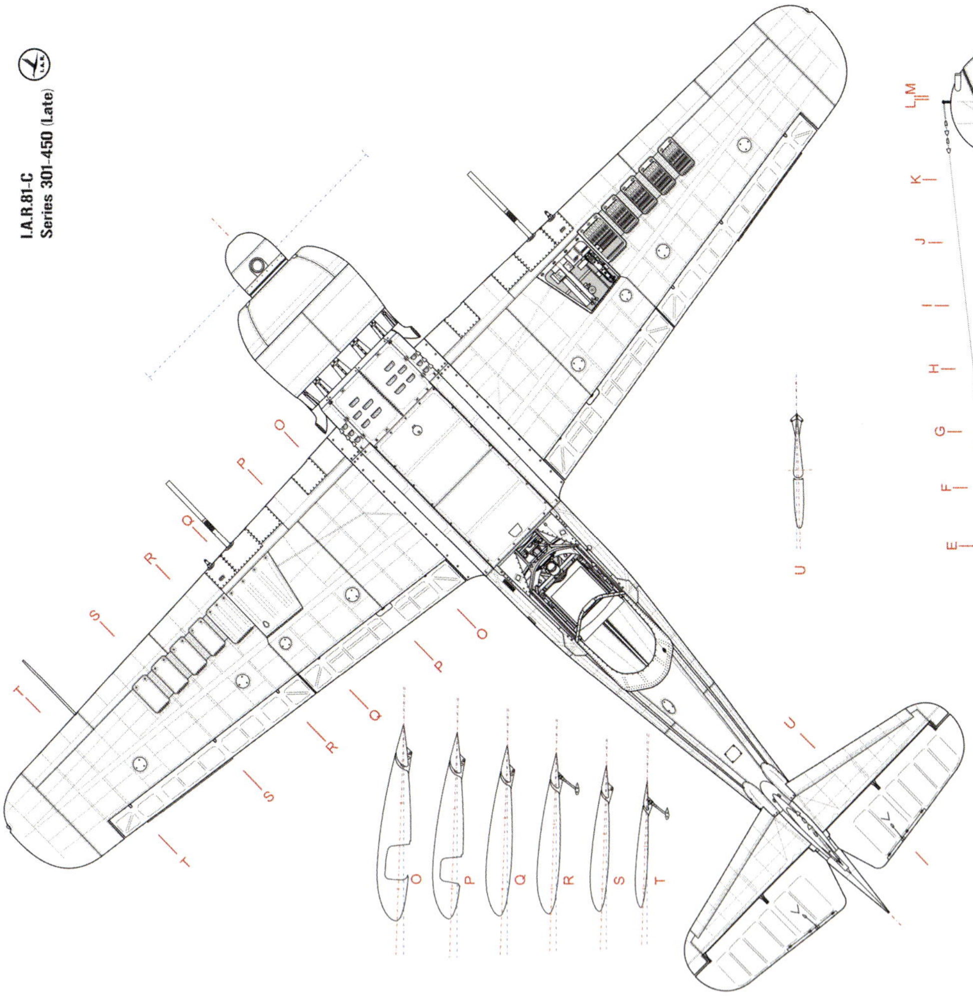

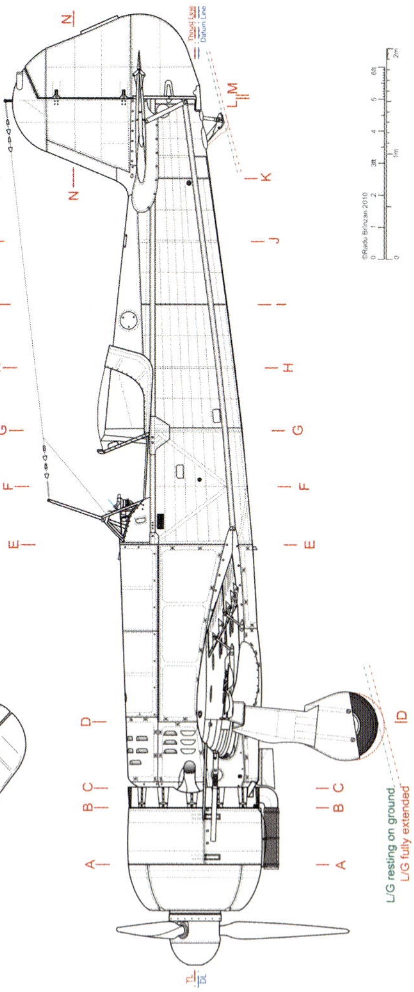

In February 1945, I.A.R.81-C No.397 was transferred to Esc.65Vt./Gr.2Vt. and took part in the Western Campaign. After the war, it served with a variety of aviation schools and in September 1955 it was recorded as 'stored' in Boboc.
Photo courtesy of ABC Collection

I.A.R.81-C No. 446 of Esc.67Vt./Gr.2Vt. was based on Gherăești Bacău airfield, during the summer of 1944. This was another aircraft flown by Adj.Av. Gheorghe Grecu on numerous escort and fighter missions on the Bessarabian front. The name 'Nina' was painted on the fuselage. The presence of the bomb-carrying equipment under the fuselage indicates that even this late into the production run, the I.A.R.81-C was still manufactured by I.A.R. Brașov with dive bombing equipment.
Photo courtesy of ABC Collection

The last of the breed! On 6 May 1944, USAF bombs that fell on an already-crippled I.A.R Brașov factory damaged or destroyed the almost-complete No.s 441, No. 442, No. 449 and No.450 and severely damaged the factory, which never resumed aircraft production. I.A.R.81-C No.448, which was subsequently assembled in Arpașul de Jos, was effectively the last aircraft ever manufactured by Regia Autonomă I.A.R. Brașov. This photograph was taken in April 1945 on Miskolc airfield in Hungary when the aircraft belonged to Esc.65Vt./Gr.2Vt. 'Note the front section of the engine cowl, which appears to be either natural metal or possibly yellow. A yellow nose section would be a very unusual feature for Romanian aircraft during the Western Campaign. If indeed, this is a yellow section, it may be a replacement part borrowed from an earlier aircraft and was awaiting painting at the time when this photo was taken.
Photo courtesy of ABC Collection

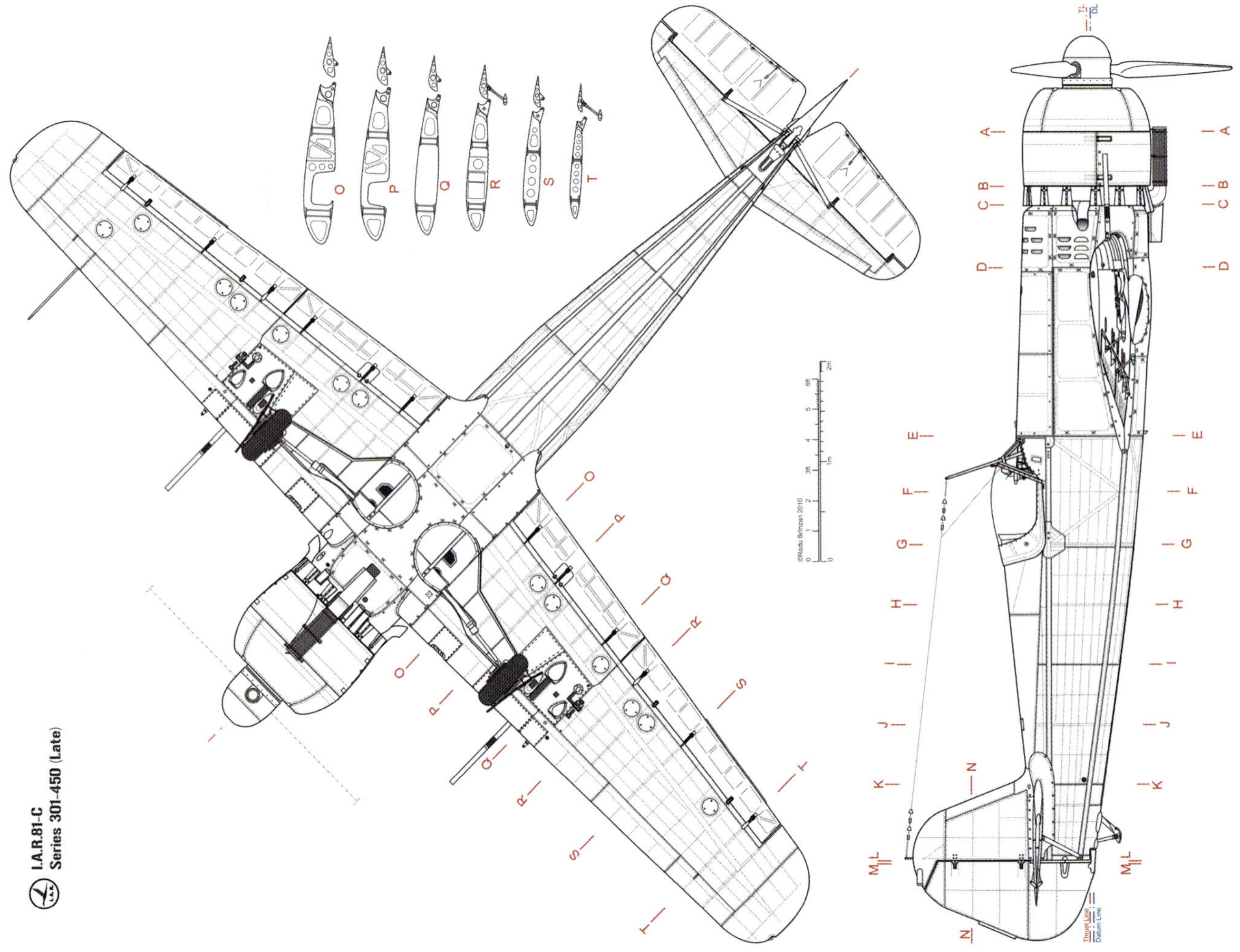

I.A.R.81-C
Series 301-450 (Late)

A group of pilots scramble towards a flight line of I.A.R.81-C on Popești-Leordeni airfield in the summer of 1944. On 10 June 1944, I.A.R.81-C No. 343 of Esc.61Vt./Gr.6Vt. flown by Lt.Av. Eugen Ianculescu was hit in the engine, propeller, cables and forward fuselage during a battle with USAF aircraft. At the time when this photo was taken, this aircraft was fitted with a special de-icing device on the supercharger intake. Note the chevron marking.
Photo courtesy of ABC Collection

At 13.25 on 4 May 1944, I.A.R.81-C No. 322 of Esc. 43Vt./Gr.7Vt. flown by Slt.Av. Mihail Vieru scrambled from Roșiorii de Vede airfield to intercept USAF aircraft. The aircraft was hit during the battle and crash-landed on the home airfield at 14.15. Note the white circle formation marking painted under the cockpit.
Photo courtesy of Pitești Military Archives via ABC Collection

On 8 August 1950, the right wing of I.A.R.81-C No. 381 of Reg.4Av. Ziliștea flown by Roman Spurcaciu hit the engine of I.A.R.81-C No. 445 while taxiing after landing on Galați airfield. The aircraft was repaired and was transferred to Reg.7Av. in Focșani, where it flew until 1952. Much earlier, this aircraft had also seen some action during the war. On 4 April 1944, when it was part of Esc.58Vt./Gr.7Vt., Cpt.Av. Horia Agarici flying this aircraft shot down one USAF B-24 Liberator. At the time when this photo was taken, this aircraft wore a yellow number 14 over the now-concealed tail band.
Photo courtesy of Pitești Military Archives via ABC Collection

On 10 June 1944, I.A.R.81-C No. 369 of Esc.62Vt./Gr.6Vt. based on Popești-Leordeni airfield flown by Lt.Av. Nicolae Limburg was shot down in a battle with USAF P-38 Lightnings and crashed near Bălăceanca killing the pilot. Note the formation markings consisting of three red triangles, the long spinner and the bomb rack.
Photo courtesy of ABC Collection

I.A.R.81-C No. 429 of Esc.67Vt./Gr.2Vt. was based on Gherăești airfield, Bacău during the summer of 1944 and it was flown by Adj.Av. Gheorghe Grecu in numerous missions escorting Junkers Ju88 bombers and Henschel Hs129 ground attack aircraft on the Bessarabian Front. The name 'Lenuța' (Little Helen) was painted on the fuselage. Note the spiral on the propeller spinner and the engine cowl with a green topside and a yellow underside. On 20 August 1944, No. 429 was lost in combat north of Iași.
Photo courtesy of ABC Collection

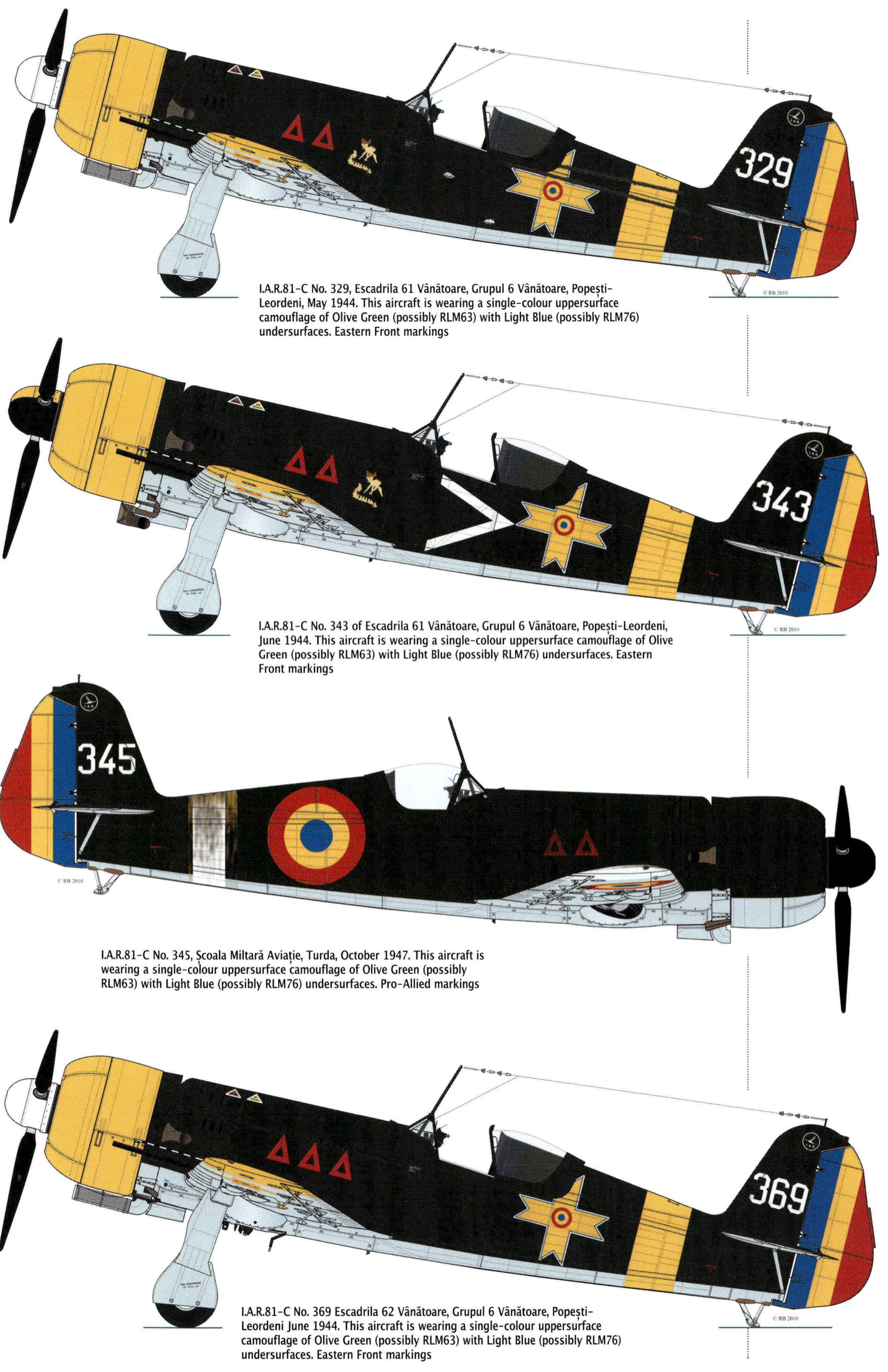

I.A.R.81-C No. 329, Escadrila 61 Vânătoare, Grupul 6 Vânătoare, Popești-Leordeni, May 1944. This aircraft is wearing a single-colour uppersurface camouflage of Olive Green (possibly RLM63) with Light Blue (possibly RLM76) undersurfaces. Eastern Front markings

I.A.R.81-C No. 343 of Escadrila 61 Vânătoare, Grupul 6 Vânătoare, Popești-Leordeni, June 1944. This aircraft is wearing a single-colour uppersurface camouflage of Olive Green (possibly RLM63) with Light Blue (possibly RLM76) undersurfaces. Eastern Front markings

I.A.R.81-C No. 345, Şcoala Militară Aviaţie, Turda, October 1947. This aircraft is wearing a single-colour uppersurface camouflage of Olive Green (possibly RLM63) with Light Blue (possibly RLM76) undersurfaces. Pro-Allied markings

I.A.R.81-C No. 369 Escadrila 62 Vânătoare, Grupul 6 Vânătoare, Popești-Leordeni June 1944. This aircraft is wearing a single-colour uppersurface camouflage of Olive Green (possibly RLM63) with Light Blue (possibly RLM76) undersurfaces. Eastern Front markings

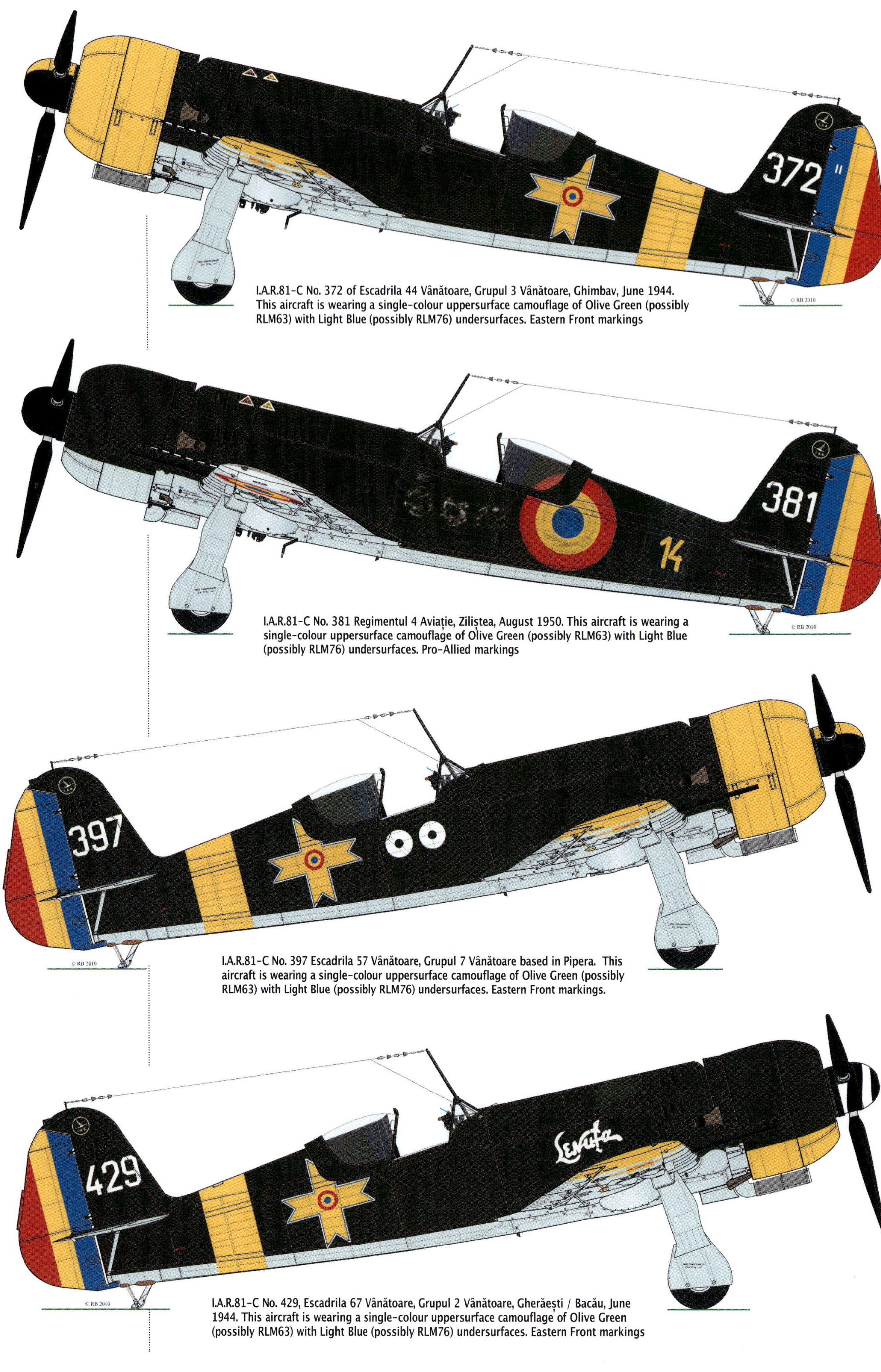

I.A.R.81-C No. 372 of Escadrila 44 Vânătoare, Grupul 3 Vânătoare, Ghimbav, June 1944. This aircraft is wearing a single-colour uppersurface camouflage of Olive Green (possibly RLM63) with Light Blue (possibly RLM76) undersurfaces. Eastern Front markings

I.A.R.81-C No. 381 Regimentul 4 Aviație, Ziliștea, August 1950. This aircraft is wearing a single-colour uppersurface camouflage of Olive Green (possibly RLM63) with Light Blue (possibly RLM76) undersurfaces. Pro-Allied markings

I.A.R.81-C No. 397 Escadrila 57 Vânătoare, Grupul 7 Vânătoare based in Pipera. This aircraft is wearing a single-colour uppersurface camouflage of Olive Green (possibly RLM63) with Light Blue (possibly RLM76) undersurfaces. Eastern Front markings.

I.A.R.81-C No. 429, Escadrila 67 Vânătoare, Grupul 2 Vânătoare, Gherăești / Bacău, June 1944. This aircraft is wearing a single-colour uppersurface camouflage of Olive Green (possibly RLM63) with Light Blue (possibly RLM76) undersurfaces. Eastern Front markings

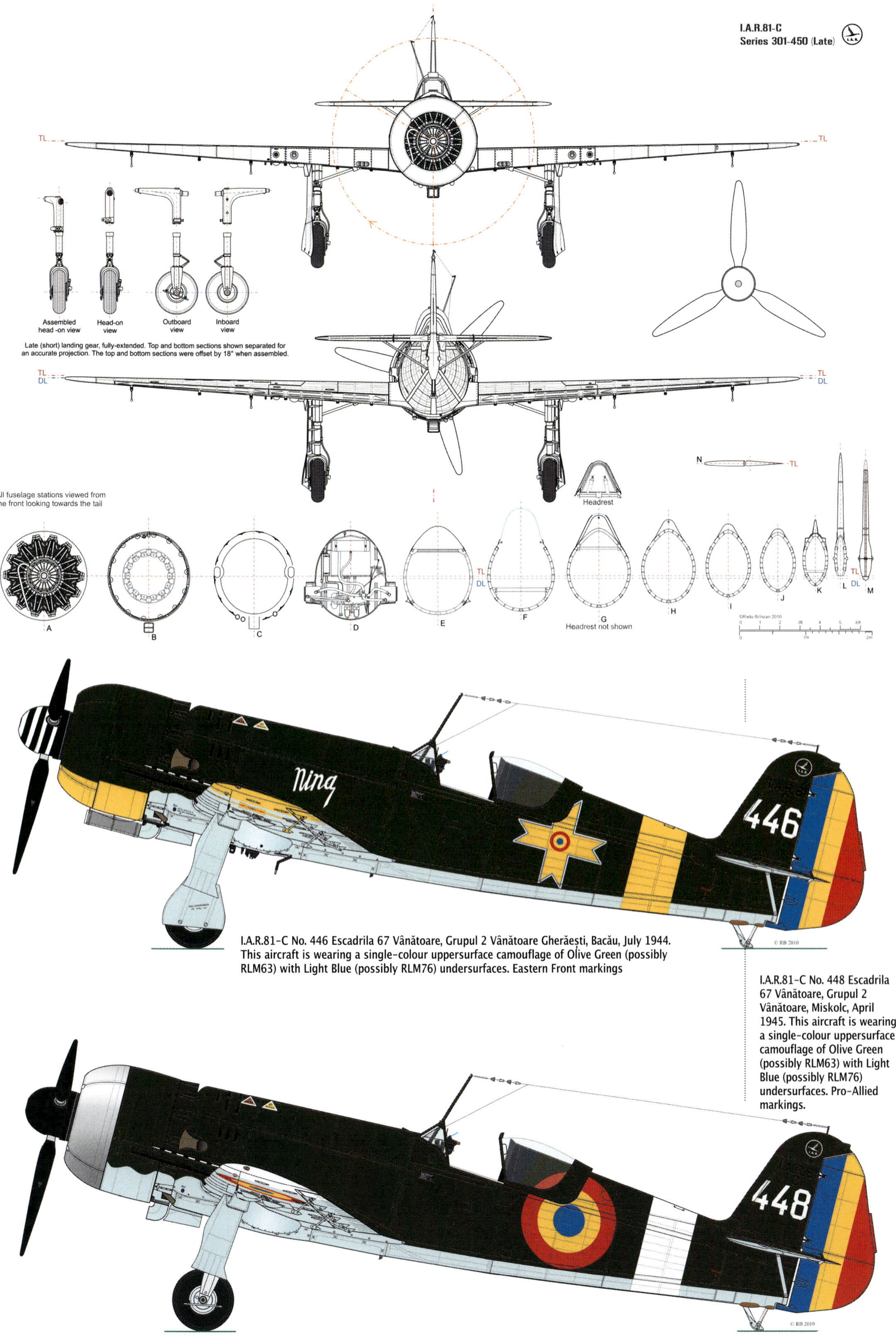

I.A.R.81-C No. 446 Escadrila 67 Vânătoare, Grupul 2 Vânătoare Gherăesti, Bacău, July 1944.
This aircraft is wearing a single-colour uppersurface camouflage of Olive Green (possibly
RLM63) with Light Blue (possibly RLM76) undersurfaces. Eastern Front markings

I.A.R.81-C No. 448 Escadrila
67 Vânătoare, Grupul 2
Vânătoare, Miskolc, April
1945. This aircraft is wearing
a single-colour uppersurface
camouflage of Olive Green
(possibly RLM63) with Light
Blue (possibly RLM76)
undersurfaces. Pro-Allied
markings.

I.A.R.80/81 modified with MG. 151/20 cannons

The so-called I.A.R.80-M

By mid-1943 it became clear that the 6-gun weapon arrangement lacked the punch to penetrate the most up-to-date armour fitted to the aircraft faced in combat by I.A.R.80/81, so a decision was taken to improve the firepower of such airframes by arming them with a pair of 20 mm Mauser MG. 151/20 cannons and a pair of 7.92 mm machine guns. Aircraft armed in this manner had the same firepower as the existing I.A.R.81-C. This re-arming programme was aimed at upgrading all six-gun airframes, especially the I.A.R.81 dive-bombers, which by that stage had been proven as inefficient in that role. The tails of some modified aircraft were marked with the letter 'M' placed under the type designation. The significance of the letter 'M' is unclear but it may stand for either for 'Mauser' or more likely for 'Modificat' (Modified). At the time of writing, it was unclear exactly how many were modified, but the intention was to eventually upgrade all six-gun aircraft. A number of airframes such as Nos.102, 103, 112 and 118 were

The emblem of Grupul 1 Vânătoare was a red 'Winged 1'.
Drawing by Teodor Liviu Moroșanu

Slt.Av. Gheorghe Gulan poses in the cockpit of I.A.R.80-M No. 104 'Getta', of Esc.64Vt., Gr.1Vt./Roșiorii de Vede Airfield in the summer of 1944. Note the late-type headrest-mounted seatbelts draped over the cockpit sill. The aircraft is still fitted with an early-type headrest. The aircraft in the background is I.A.R.80-M No. 169.
Photo courtesy of ABC Collection

Slt.Av. Gheorghe Gulan and two other airmen look at I.A.R.80-M No. 104 "Getta", of Esc.64Vt./Gr.1Vt., Roșiorii de Vede Airfield in the summer of 1944. Note the three-bar formation marking, the early-type headrest fitted with late-type seatbelts as well as the parachute on the seat
Courtesy of ABC Collection

sent to be modified but the upgrade work was not completed for various reasons, such as shortage of supplies or enemy attack on the premises. The following 6-gun aircraft are known to have been upgraded with MG. 151/20 cannons: 91, 92, 93, 94, 96, 97, 104, 108, 110, 111, 118, 120, 122, 123, 125, 128, 129, 133, 141, 151, 153, 159, 162, 163, 165, 166, 169, 170, 173, and 174, but this is not a definitive list.

I.A.R.80-M No. 111 of Esc.65Vt./Gr.2Vt. is undergoing a technical inspection on Zvolen airfield in April 1945 during the Western Campaign. Interestingly, the tail of this aircraft seems to have an unusual mark after the serial number. This may be either a patch of chipped paint or the inscription 'B2'.
Photo courtesy of ABC Collection

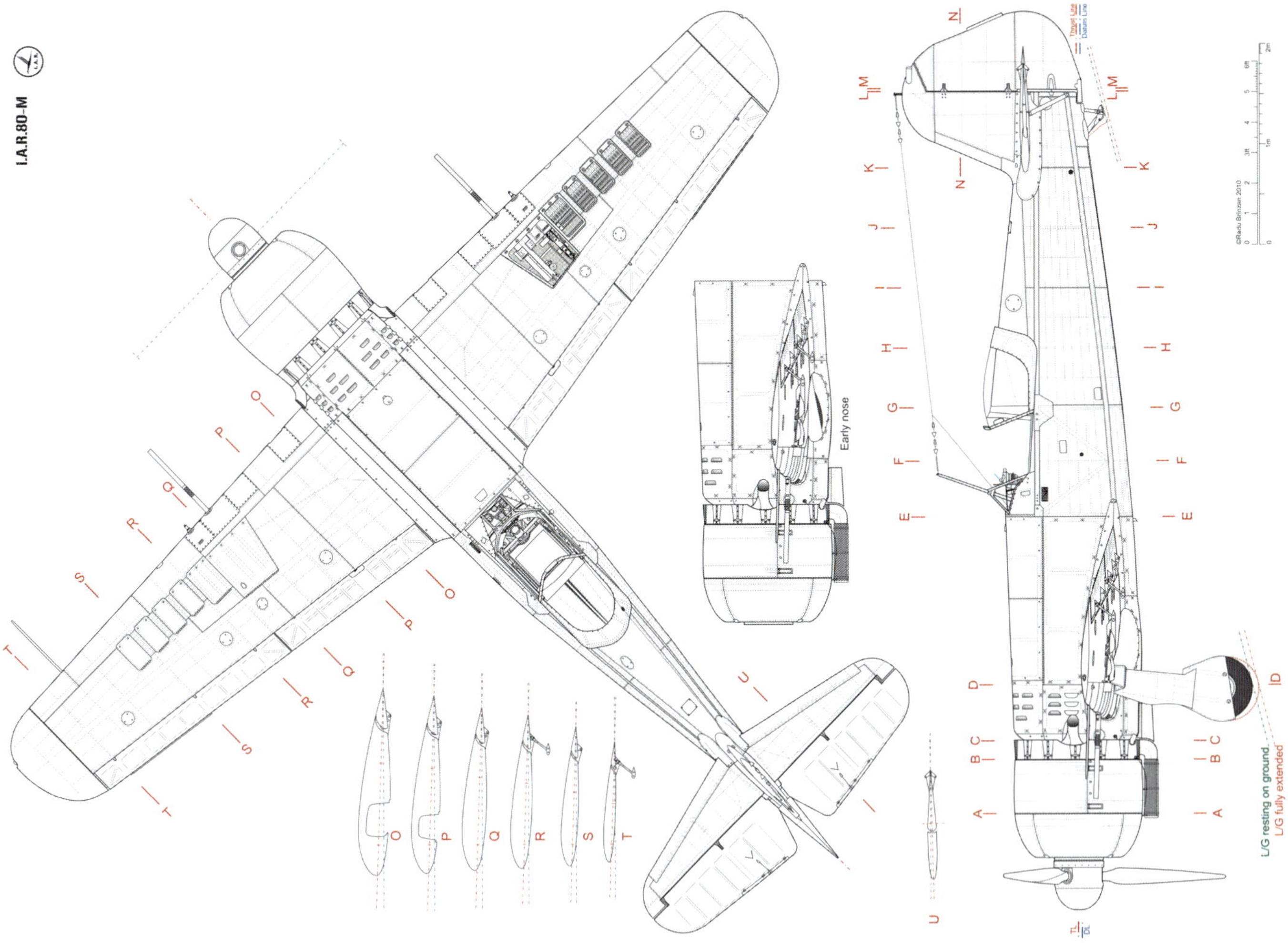

Early nose

L/G resting on ground.
L/G fully extended

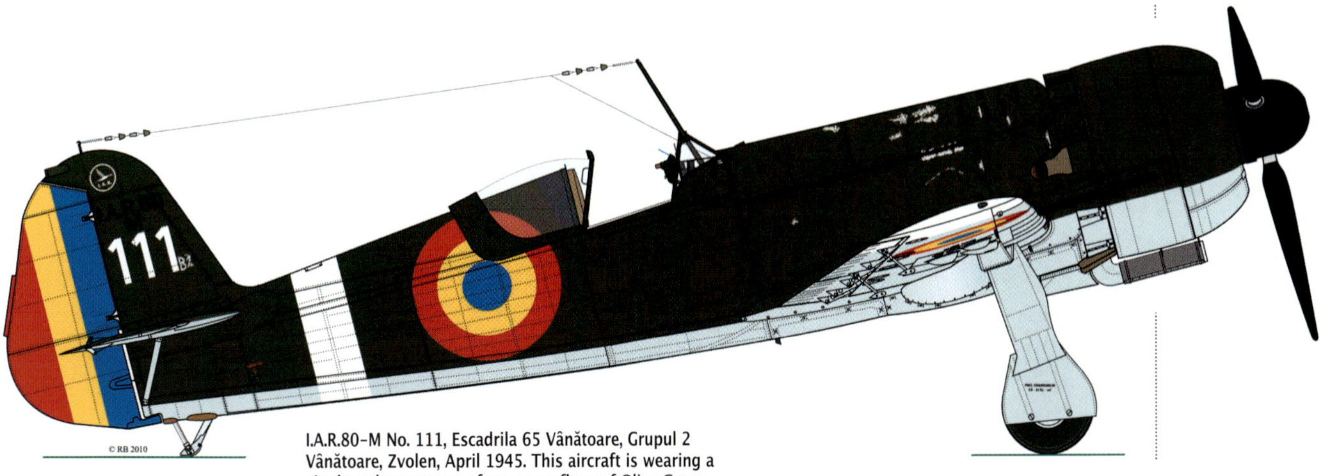

©Radu Brinzan 2010

I.A.R.80-M No. 111, Escadrila 65 Vânătoare, Grupul 2 Vânătoare, Zvolen, April 1945. This aircraft is wearing a single-colour uppersurface camouflage of Olive Green (possibly RLM64) with Light Blue (possibly RLM76) undersurfaces. Pro-Allied markings

© RB 2010

This photograph shows I.A.R.80-M No. 169 of Esc.64Vt./Gr.1Vt. based on Roșiorii de Vede airfield in June 1944. Note the Gr.1Vt. emblem on the nose and the three-bar formation markings. On 24 June 1944, while flown by Slt.Av. Petre Mihăilescu this aircraft took part in an aerial battle with USAF aircraft in which the aircraft was hit in the wing, control surfaces and fuselage. The aircraft was repaired and flew again.
Photo courtesy of ABC Collection

Adj.Av. Vasile Stana leans on his personalised I.A.R.80-M No.151 of Esc.46Vt./Gr4.Vt., in August 1944. On 10 August 1944, while he and his wingman were escorting Savoia SM-79 bombers over Bessarabia, they were attacked by four Soviet P-39 Airacobras. In the ensuing battle, Adj.Av. Stana shot down one P-39 Airacobra near the locality of Soroca.
Photo courtesy of ABC Collection

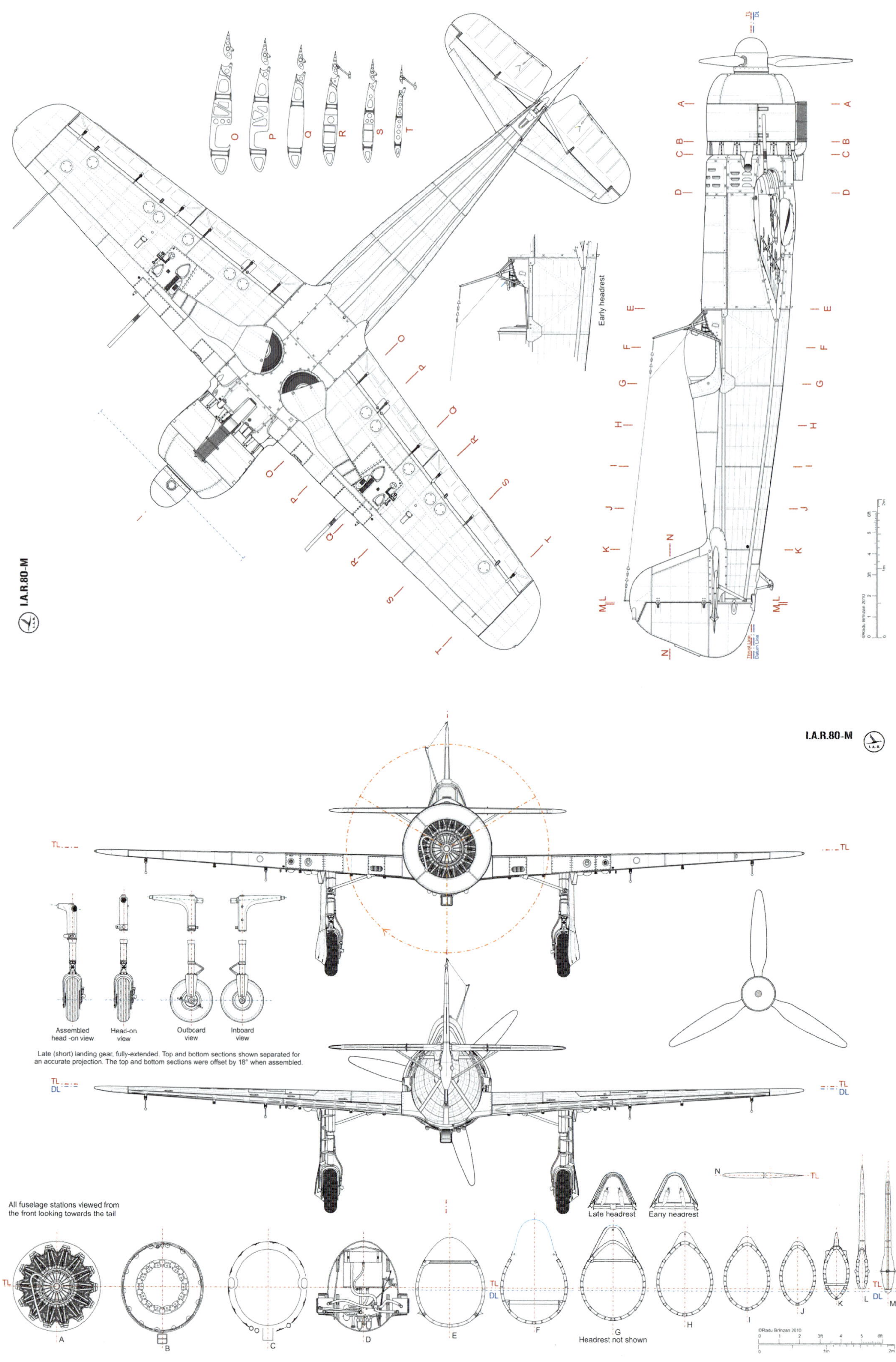

I.A.R.80-M
I.A.R.80-M
Early headrest
©Radu Brinzan 2010
Assembled head-on view
Head-on view
Outboard view
Inboard view
Late (short) landing gear, fully-extended. Top and bottom sections shown separated for an accurate projection. The top and bottom sections were offset by 18° when assembled.
All fuselage stations viewed from the front looking towards the tail
Late headrest
Early headrest
Headrest not shown
©Radu Brinzan 2010

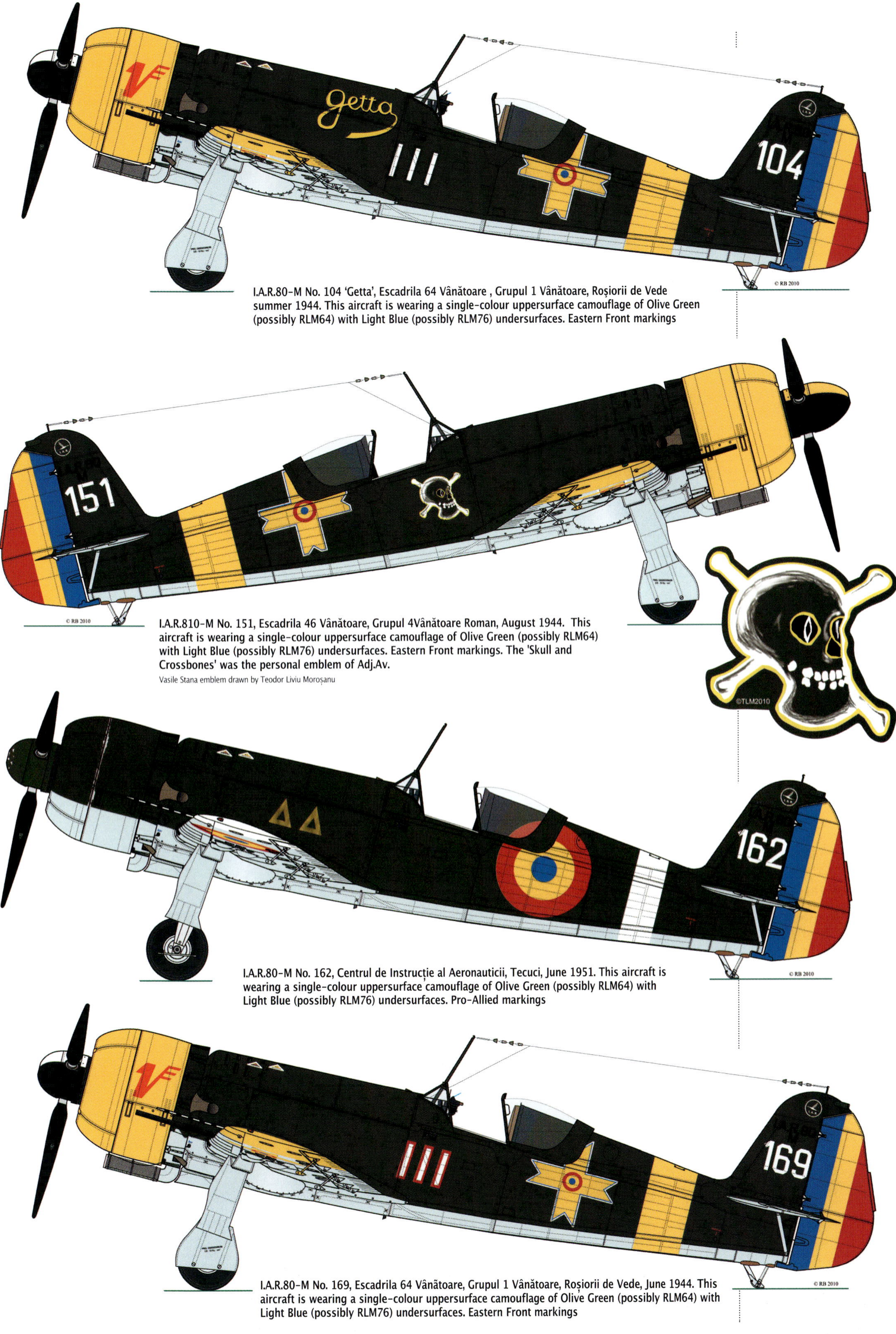

I.A.R.80-M No. 104 'Getta', Escadrila 64 Vânătoare , Grupul 1 Vânătoare, Roşiorii de Vede summer 1944. This aircraft is wearing a single-colour uppersurface camouflage of Olive Green (possibly RLM64) with Light Blue (possibly RLM76) undersurfaces. Eastern Front markings

I.A.R.810-M No. 151, Escadrila 46 Vânătoare, Grupul 4Vânătoare Roman, August 1944. This aircraft is wearing a single-colour uppersurface camouflage of Olive Green (possibly RLM64) with Light Blue (possibly RLM76) undersurfaces. Eastern Front markings. The 'Skull and Crossbones' was the personal emblem of Adj.Av.

Vasile Stana emblem drawn by Teodor Liviu Moroşanu

I.A.R.80-M No. 162, Centrul de Instrucţie al Aeronauticii, Tecuci, June 1951. This aircraft is wearing a single-colour uppersurface camouflage of Olive Green (possibly RLM64) with Light Blue (possibly RLM76) undersurfaces. Pro-Allied markings

I.A.R.80-M No. 169, Escadrila 64 Vânătoare, Grupul 1 Vânătoare, Roşiorii de Vede, June 1944. This aircraft is wearing a single-colour uppersurface camouflage of Olive Green (possibly RLM64) with Light Blue (possibly RLM76) undersurfaces. Eastern Front markings

Two-seat trainer I.A.R.80-DC

On 8 December 1948, Comandamentul Aeronauticii Militare, Direcţia Tehnică proposed that a a number of I.A.R.80/81 be modified into I.A.R.80-DC training aircraft. The suffix 'DC' is the acronym for Dublă Comandă (Twin Control). The aircraft were modified during 1950-1951 by ASAM Pipera. The first airframe modified was No.447 which joined Şcoala Militară de Aviaţie in Focşani in 1950. A second cockpit was installed in the fuselage, ahead of the existing cockpit. The forward cockpit was open but fitted with a windscreen. This cockpit was located in the place previously occupied by the fuel tank and the radio. A new smaller tank had to be fitted and the range of the aircraft was significantly reduced. However, the trainer aircraft retained the series aircraft's capability to carry drop tanks, which could be used to increase range. There is no information about the type of radio fitted to these aircraft, but it is very unlikely that it was the obsolete, heavy and sizeable FuG VIIa set, which by this stage was most likely replaced by a newer and lighter set, probably a Soviet model. Because the majority of DC aircraft were based on the I.A.R.81-C, they could carry Mauser MG150/20 cannons and/or Browning FN 7.92 machine guns for training purposes, but most likely they did not carry armament in order to reduce weight. It is unclear whether earlier-type aircraft were upgraded with the type of wings that could carry the Mauser cannons. The pilots and instructors were quite pleased with the performance of this trainer aircraft. The aircraft known to have been modified were Nos. 179, 188, 190, 230, 245, 302, 319, 329, 364, 378, 383, 408, 410, 425, 436, and 447, but this list is not definitive.

I.A.R.80-DC No. 425 of Şcoala Militară Aviaţie Tecuci, April 1951.
This aircraft is wearing a single-colour uppersurface camouflage
of Olive Green (possibly RLM64) with Light Blue (possibly RLM76)
undersurfaces. Pro-Allied markings.

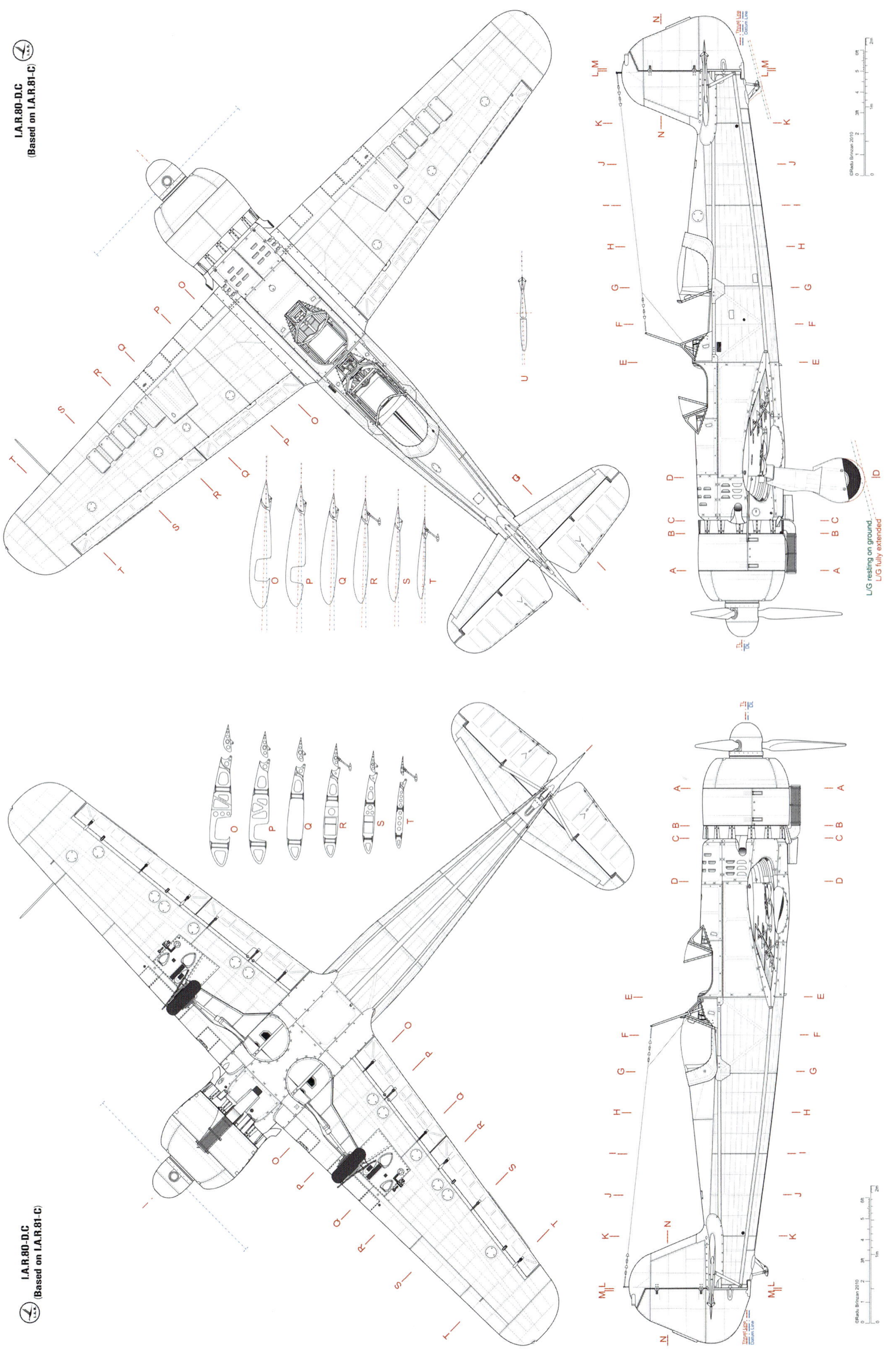

I.A.R.80-DC
(Based on I.A.R.81-C)
L/G resting on ground
L/G fully extended
Two-seat trainer I.A.R.80-DC
73

Experimental I.A.R.80 with In-line Engine

By the end of 1940, it became evident that the 1,000 hp I.A.R.-14K engine was superseded by the engines fitted to the existing fighter aircraft. In December 1940, MAM requested that the 1175 hp Daimler Benz DB-610A engine be tested on the I.A.R.80. For this purpose, Messerschmitt Bf.109-E No.4 was delivered to I.A.R. Braşov where its engine was removed and fitted to I.A.R.80 No. 13. This was test-flown on 21 April 1941 by Ing. Alexandru Frim, but the vibration caused by the engine nearly caused the loss of the aircraft, which was brought back to land with great difficulty. As a result, the project was abandoned. In 1943, I.A.R. Braşov decided to purchase a number of 1,475 hp DB-605 engines from Germany that were intended for use on the licence-built Messerschmitt Bf.109 aircraft that the factory was due to start manufacturing soon. In light of the expected availability of this engine, I.A.R. Braşov decided to test it with the I.A.R.80/81. One trial was carried out with I.A.R.81-C No. 326 that flew successfully between 29 June and 20 August 1944. The reports stated that the performance improved noticeably and MAM took the decision to equip the I.A.R. 81 with the DB605 engine. In the end, due to shortages of engine supplies from Germany where these engines were in great demand, this decision was never implemented.

This is the only known photo of an I.A.R.80/81 equipped with an in-line engine. The exact identity of this aircraft or the date when this photograph was taken is not known. The engine cowl, spinner and coolant radiator appear to be borrowed from a Savoia Marchetti SM-79 JRS B bomber. A Venturi tube was mounted below the windscreen next to the antenna lead-in insulator. The presence of the King Michael Cross indicates that this photo was taken at some stage after the end of May 1941. The cross is of an early type with thick blue lines. The wing features no visible armament, but it may be a four-gun wing as indicated by the absence of outboard armament and the associated spent ammunition chutes or may be a Mauser-cannon wing with all armament removed. The windscreen appears to be an early type without an internal reinforced armoured glass panel.
Courtesy of ABC Collection

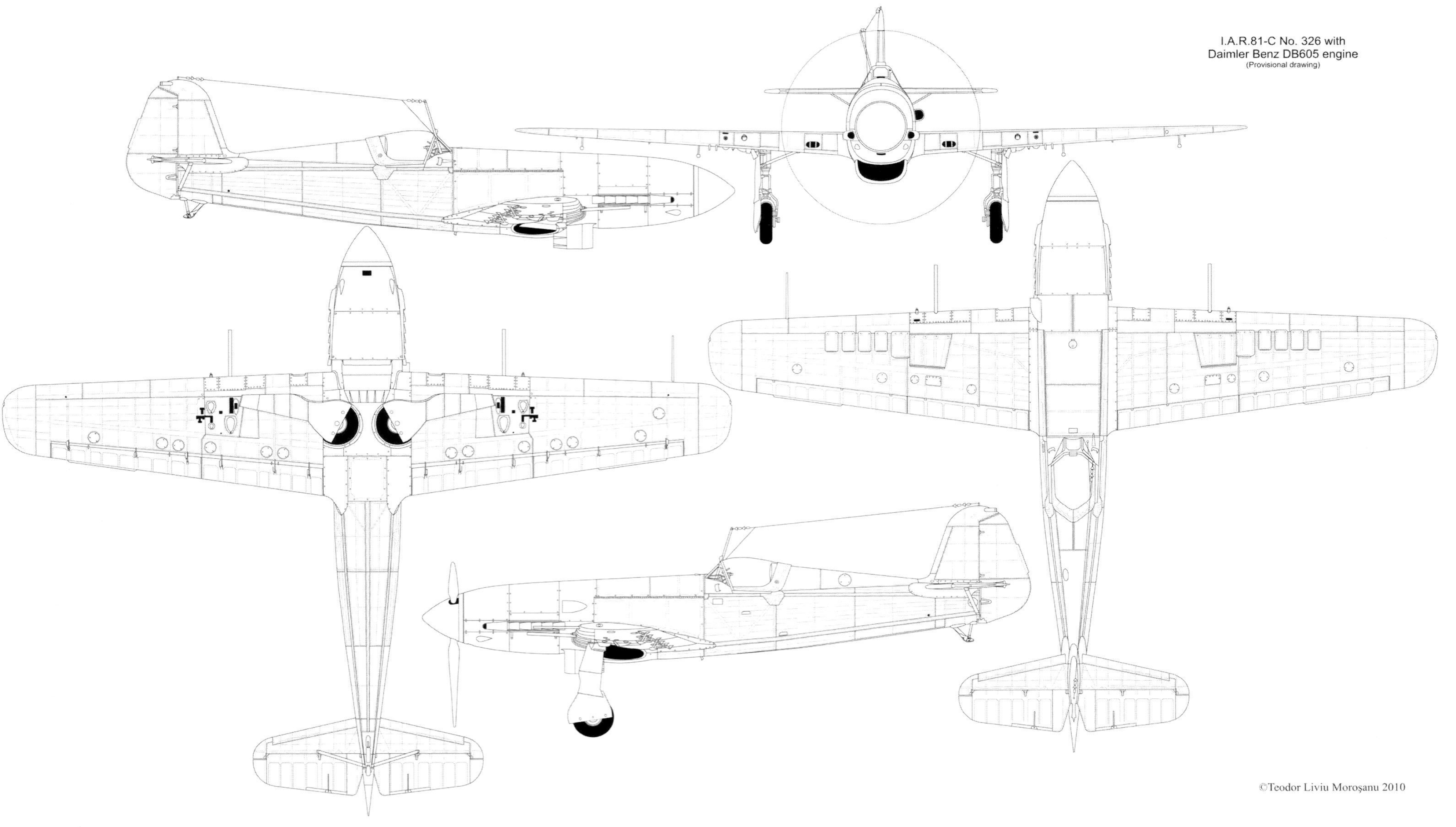

I.A.R.81-C No. 326 with
Daimler Benz DB605 engine
(Provisional drawing)
©Teodor Liviu Moroşanu 2010

Evolution of the I.A.R.80 & I.A.R.81

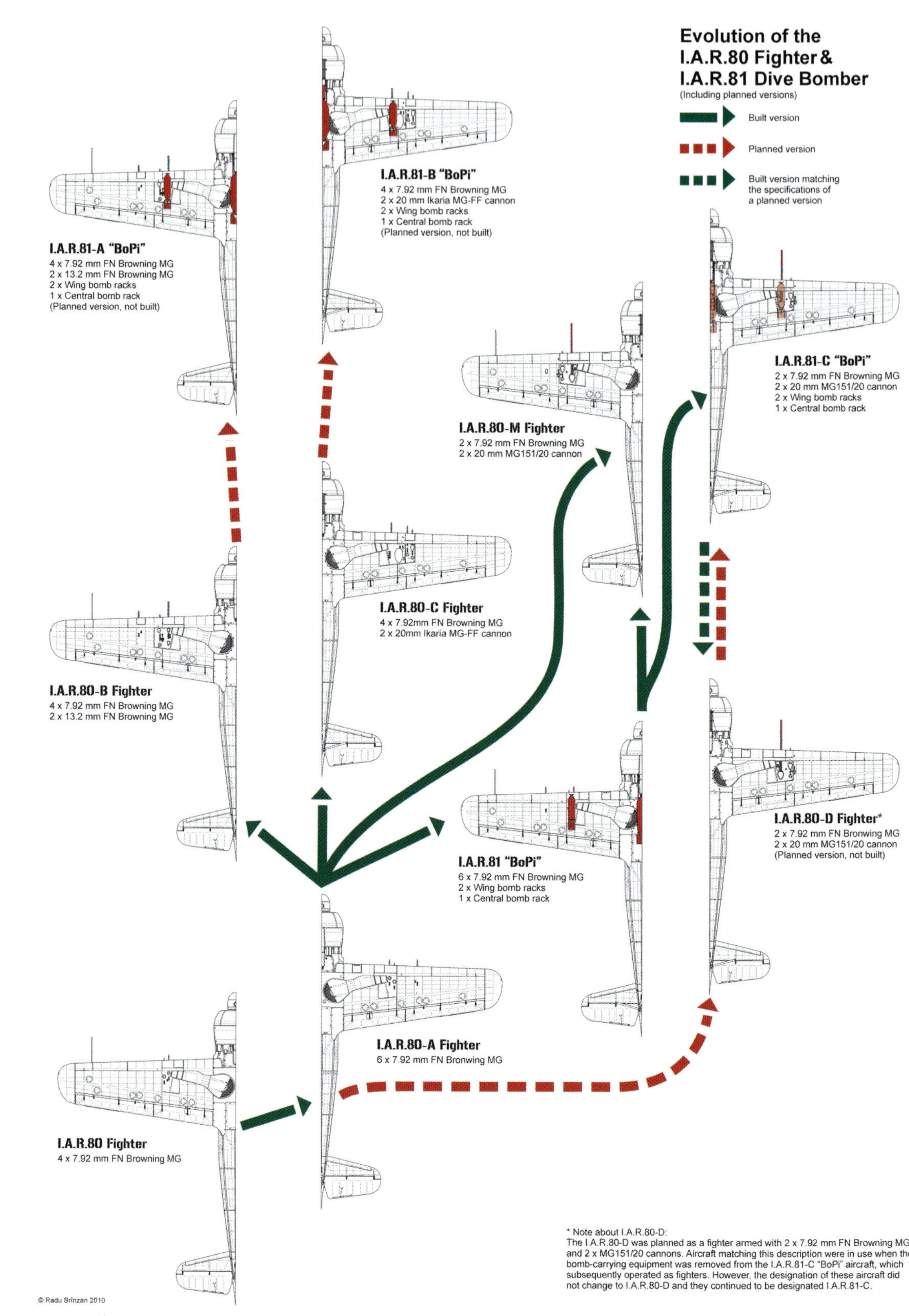

Designation		I.A.R.80	I.A.R.80	I.A.R.80-A	I.A.R.80-B			I.A.R.80-C	I.A.R.81			I.A.R.80-B	I.A.R.81-C
Type		Prototype	Fighter	Fighter	Fighter			Fighter	Dive Bomber			Fighter	Dive Bomber**
Measurements and performance	Unit of measurement	Serial Numbers											
		0	1- 50	51 - 75 76 - 90 106 - 150 176 - 180	181 - 200	201 - 211	212 - 230	241 - 290	91 - 105	151 - 175	231 - 240	291 - 300	301 - 450
Wingspan	mm	10,000	10,520	10,520	10,520	10,520	11,000	11,000	10,520	10,520	10,520	11,000	11,000
Length	mm	8,160	8,900	8,900	8,900	8,970	8,970	8,970	8,900	8,900	8,970	8,970	8,970
Height	mm	3,600	3,600	3,525*	3,525	3,525	3,525	3,525	3,525	3,525	3,525	3,525	3,525
Lift area	Sq. mm	15,500	16,000	16,000	16,000	16,000	16,500	16,500	16,000	16,000	16,000	16,500	16,500
Empty weight	kg	1,780	2,045	2,110	2,135	2,135	2,190	2,200	2,155	2,155	2,155	2,190	2,200
Maximum weight, fighter	kg	2,270	2,483	2,720	2,810	2,810	2,870	2,880	2,750	2,780	2,800	2,870	2,900
Maximum weight, with bombs	kg	-	-	-	-	-	-	-	3,070	3,100	3,125	-	-
Maximum weight, with drop tanks	kg	-	-	-	-	2,960	3,020	3,030	-	-	2,920	3,020	3,060
Fuel load	kg	330	330	330	330	330+146	330+146	330+146	330	330	330+146	330+146	330+146
Operational ceiling	m	11,000	10,500	10,500	10,000	10,000	10,000	10,000	10,000	10,000	10,000	10,000	10,000
Climb to 5,000 m, fighter	Min./sec.	-	6'0''	5'27''	5'50''	6'30''	6'30''	6'45''	7'00''	5'50''	6'30''	6'30''	7'00'
Climb to 5,000 m, with bombs	Min./sec.	-	-	-	-	-	-	-	7'30''	7'30''	7'30''	-	-
Max. speed at 5,000 m, fighter	km/h	510	485	485	485	485	485	485	485	485	485	485	485
Max. speed at 5,000 m, with bombs	km/h	-	-	-	-	-	-	-	465	465	465	-	-
Max. speed at 5,000 m, with drop tanks	km/h	-	-	-	-	465	465	465	465	465	465	465	465
Minimum speed	km/h	200	200	205	205	205	205	205	215	215	215	205	210
Landing speed	km/h	170	170	175	175	175	175	175	175	175	180	175	180
Take-off distance, fighter	m	260	260	300	300	300	300	300	300	350	350	300	400
Take-off distance, with bombs	m	-	-	-	-	-	-	-	500	500	500	-	-
Landing distance	m	300	350	350	400	400	400	400	350	350	400	400	400
Range, fighter	km	700	760	730	730	730	730	730	730	730	730	730	730
Range, with bombs	km	-	-	-	-	-	-	-	695	695	695	-	-
Range, with drop tanks	km	-	-	-	-	1,030	1,030	1,330	1,330	1,330	1,330	1,030	1,330
Armament and external payload		2 x 7.92 mm FN Browning MG	4 x 7.92 mm FN Browning MG	6 x 7.92 mm FN Browning MG	4 x 7.92 mm FN Browning MG 2 x 13.2 mm FN Browning MG	4 x 7.92 mm FN Browning MG 2 x 13.2 mm FN Browning MG 2 x 50 kg wing - mounted drop tank racks	4 x 7.92 mm FN Browning MG 2 x 13.2 mm FN Browning MG 2 x 50 kg wing - mounted drop tank racks	4 x 7.92 mm FN Browning MG 2 x 20 mm Ikaria MG/FF cannon 2 x 100 kg wing - mounted drop tank racks	6 x 7.92 mm FN Browning MG 1 x 250 kg central bomb rack 2 x 50 kg wing - mounted bomb racks	6 x 7.92 mm FN Browning MG 1 x 250 kg central bomb rack 2 x 50 kg wing - mounted bomb racks	6 x 7.92 mm FN Browning MG 1 x 250 kg central bomb rack 2 x 50 kg wing - mounted bomb/drop tank racks	4 x 7.92 mm FN Browning MG 2 x 13.2 mm FN Browning MG 2 x 100 kg wing - mounted drop tank racks	2 x 7.92 mm FN Browning MG 2 x 20 mm Mauser M G 151/20 cannon 1 x 250 kg central bomb rack 2 x 100 kg wing - mounted drop tank/bomb racks
Remarks		*Height (aircraft o n the ground): - 3,600 mm starting with aircraft No. 1 until aircraft No. 75. - 3,525 mm starting with aircraft No. 76 until the end of production. **Although the I.A.R.81-C left the factory as a dive bomber, it was effectively a fighter. The drag-cau sing central bomb rack on I.A.R.81-C was usually removed in the field. The wing mounted racks were used to carry drop tanks.											

3.1-1 I.A.R.80 Left Side (Viewed from the right)

1. Air lines departing from the auxiliary compressor
2. Lines that supply air to the hydraulic pump
3. Flexible support for the fuel tank
4. Throttle control linkages
5. Hydraulic storage tank for the landing gear
6. Oxygen bottle for the breathing apparatus
7. Hydraulic storage tank for the flaps
8. Air filter
9. Flow regulator for the breathing apparatus
10. Throttle lever
11. Rear-view mirror
12. Fuel cut-off controls
13. Landing gear distributor controls
14. Tube linking the flow regulator with the breathing mask
15. Map case
16. Seat support lever
17. Electrical socket
18. Elevator trim control lever
19. Elevator trim tab
20. Elevator trim control cables
21. Elevator trim control wheel
22. Clutch selector for the engine pump (hydraulic)
23. Wing flap distributor control
24. 30/8 kg/cm2 pressure reducer
25. Distributor relay
26. Couplings
27. Selectors for the lines leading to the machine guns
28. Air lines for the bottle pressure gauge
29. Outlet for the oil gathered in the filter
30. Air lines for firing the machine guns
31. Axle and lever controlling the ailerons
32. Wing cross-beam
33. Brake lines
34. Connection point for the landing gear hydraulic ram
35. Hook securing the landing gear in the retracted position
36. Brake control pistons
37. Wheel yoke
38. Landing gear door
39. Throttle control rods
40. Intake limiter control
41. Fuel cut-off control
42. Hydraulic pump
43. Lateral exhaust outlet
44. Lower exhaust outlet
45. Engine bearer
46. Carburettor
47. Gear box breather tube joining the exhaust pipe. [Starting
    with No. 91, this pipe was widened and extended to
    discharge under the fuselage, ahead of the cockpit. See
    cutaway drawing for I.A.R.81]
48. Choke handle
49. D.B.U. group valves
50. Controls for the 'Caretta' valve (supplementary tank)
51. Control rod for the 'Caretta' valve
52. 'Caretta' valve

Original factory drawing courtesy of Dan Antoniu

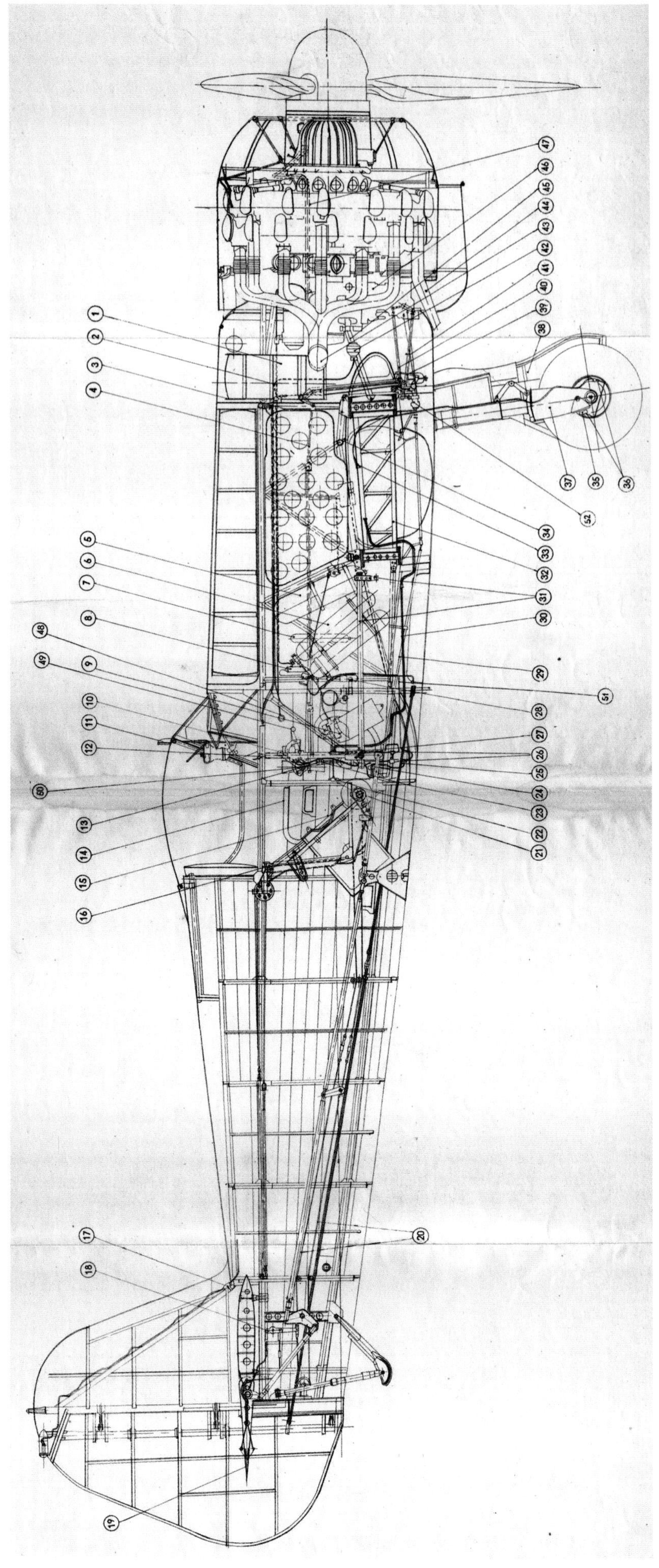

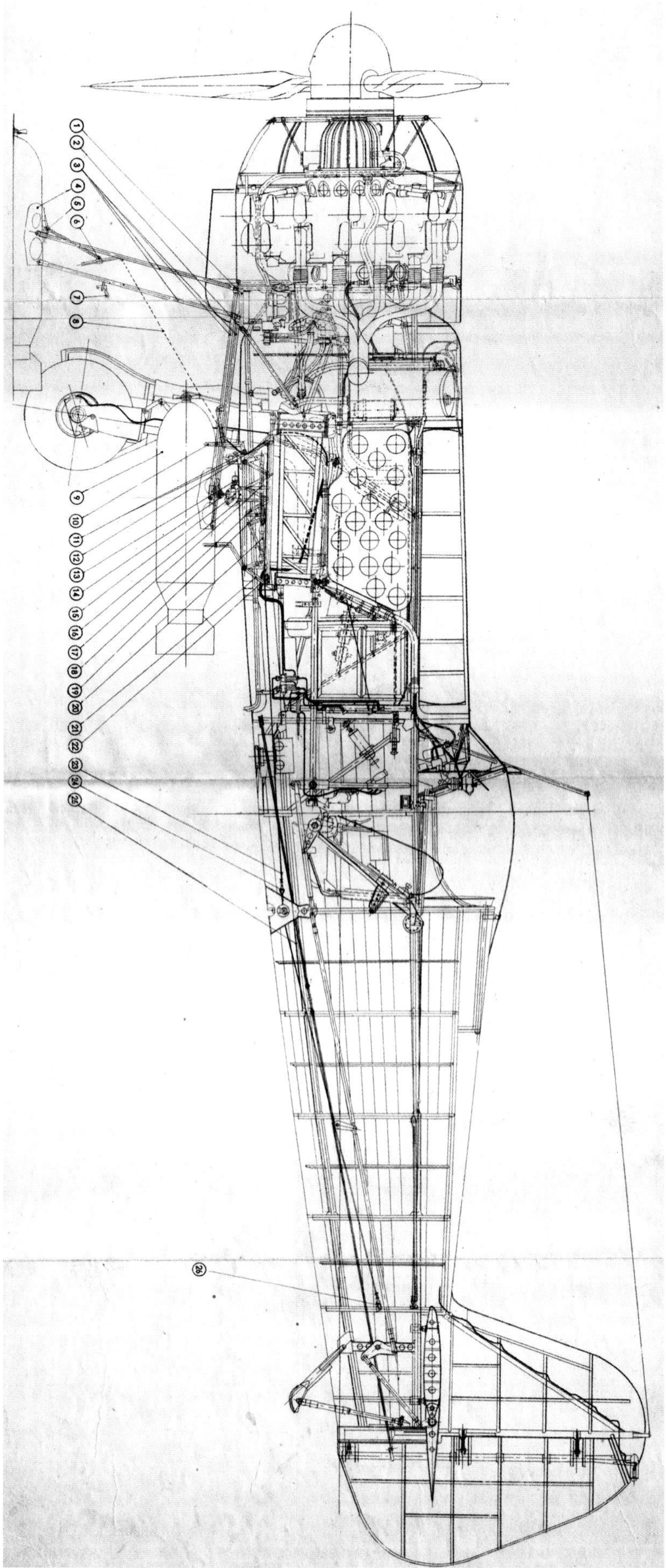

3.1-2 I.A.R.81 Right Side
(Viewed from the left)

1. Pyramid
2. Pyramid cross-brace
3. Trapeze launcher (in normal and launching position)
4. Saddle of the trapeze launcher
5. Sleeve
6. Deployment device
7. Attachment hook
8. Trapeze launcher retraction cable
9. 225 kg bomb
10. Bomb sway-brace
11. Pulleys
12. Bomb hook housing
13. Thermo-electrical device
14. Standard bomb attachment lug
15. Pneumatic device
16. Lock
17. Linkage rod for bomb sway-brace adjustment
18. Threaded rod for bomb sway-brace adjustment
19. Slide for bomb sway-brace adjustment
20. Linkage rod for bomb sway-brace adjustment
21. Bomb carrier body
22. Pulley
23. Cable
24. Eyelet anchor
25. Trapeze launcher retraction bungee cord
26. Eyelet anchor
Original factory drawing courtesy of Dan Antoniu

Forward & Rear Fuselage

Forward Fuselage

A firewall was attached to the front end of the fuselage and separated the engine from the fuel tank. The large tank attached to it is the oil tank and the small black cylinder next to it is the motor for the propeller pitch controller. The diagonal bar is the engine bearer.
Photo from the technical manual courtesy of ABC Collection

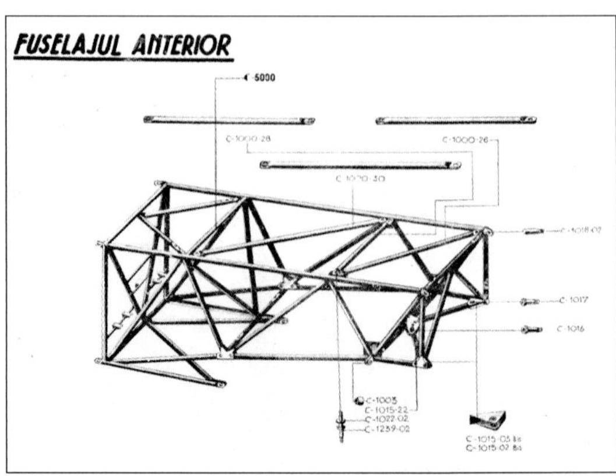

This is a perspective drawing of the internal frame of the forward fuselage. The two diagonal cross braces and the horizontal cross-bar at the top are removable to allow for the removal of the fuel tank. This structure was the backbone of the aircraft, to which were attached the rear fuselage at the back, the engine bearer at the front, the wing underneath, the fuel tanks inside and above, the radio set at the rear and the retraction mechanism of the landing gear.
Illustration from the parts list courtesy of ABC Collection

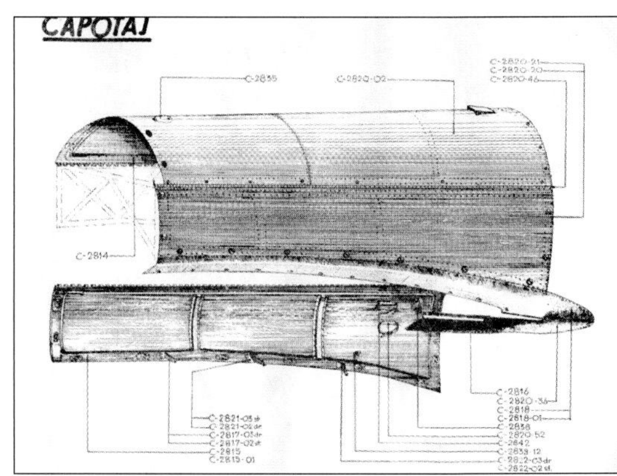

The forward fuselage panels.
Illustration from the parts list courtesy of ABC Collection

This picture provides a closer look at the various devices accessible by opening the left panel. a = Oxygen bottle for the breathing apparatus. The large bottle on the left is the hydraulic tank for the landing gear. The rectangular box barely visible at the bottom is the battery. The large item that occupies almost the entire left half of the opened space is the lower fuel tank.
Photo from the technical manual courtesy of ABC Collection

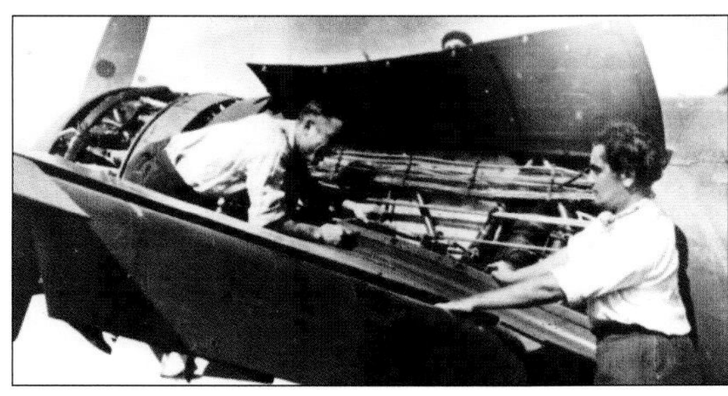

The side panels of the forward fuselage were hinged and could be opened for easy servicing. This allowed access to engine controls and various pressurised vessels such as hydraulic tanks, compressed air tank and breathing oxygen tank. From this side it was also possible to access the battery, which is located below the right elbow of the man standing in this picture. The similar panel on the opposite side allowed access to the radio set. This panel was held in place by six fasteners that can be seen on the edge of the open panel. This aircraft is I.A.R.81-C No. 446 photographed on Gherăeşti airfield in the summer of 1944.
Photo courtesy of ABC Collection

A - Short nose, used up to I.A.R.80 No. 75 (long landing gear)
B - Long nose, used starting with No. 201 (extended by 70 mm)
C - DC nose (trainer aircraft)
1 - Cooling slots
2 - Access panel to throttle controls
3 - Compress air fill point
4 - Wing fairing
5 - Fuel fill hatch
6 - Cockpit ventilation scoop
7 - Hinged access panel to oil tank (oil fill)
8 - Short side panel
9 - Electrical socket
10 - Outlet for the oil cooler
11 - Engine starter crank
12 - Opening for the landing gear up-lock
13 - Access panel to fuel drain
14 - Central panel used on aircraft with long landing gear
15 - Long side panel with cooling gills - used on aircraft with a second oil cooler in the left wing.
16 - Bar with mounting points for the trapeze bomb launcher (used only on I.A.R.81 and I.A.R.81-C)
17 - Central panel used on I.A.R.81 with slots for the bomb rack
18 - Bomb rack (used only on I.A.R.81 and I.A.R.81-C)
19 - Central panel used on aircraft with short landing gear
20 - Blister fairing over the connections for the second oil cooler
21 - Long side panel without cooling gills
22 - Indentation for the exhaust pipe
23 - Windscreen
24 - Outlet of the engine gear box breather tube mechanism (used starting with No. 91)
25 - Reinforcement strap (used from No. 20 to No. 93)
26 - Service panel for the emergency canopy release device (starting with No. 241 and retrofitted to some existing airframes)
27 - Step (left side only)
28 - Signal flare outlet
29 - Service panel for rudder cables
30 - Fairing over the rudder cable control horn
31 - White position light

I.A.R.80 and I.A.R.81
Various fuselage details and panels

32 - Tail plane support strut (used on I.A.R.80-C, I.A.R.81 and I.A.R.81-C)
33 - Roof of the wheel well
34 - Strengthening straps
35 - Tail plane fairing
36 - Antenna wire
37 - Canopy lock handle
38 - Antenna lead-in insulator
39 - Zipped service access to trim tab cable
40 - Trim tab
41 - Flared fairing in the cockpit roof to clear the landing gear retraction ram
42 - Access window for the landing gear retraction ram couplings
43 - Hole for hoist bar
44 - Fairing for engine bearer
45 - Opening in the underside for the compressor intake
46 - Fairing for the oil cooler outlet
47 - Drain tube for oil cooler
48 - Internal structure of elevator

©Radu Brinzan 2010

Rear Fuselage

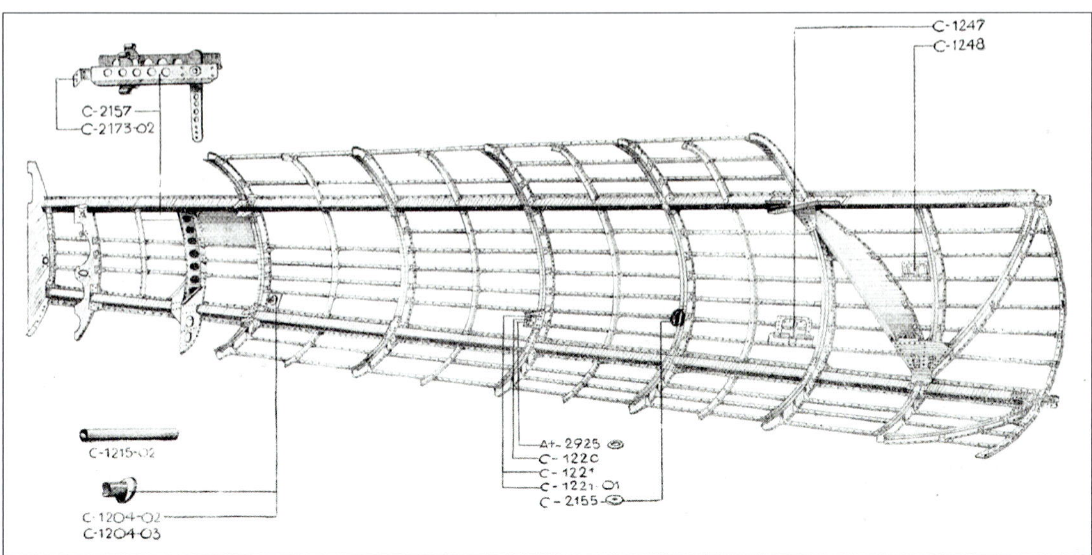

The left side of the rear fuselage viewed from inside (from the right). The rear fuselage was a semi-monocoque consisting of nine main frames with a double-T cross-section, five intermediate frames with a Z cross-section through which passed four main longerons with a flanged-U cross-section. The front ends of these longerons were attached to the rear of the internal nose frame. The space between the first three frames was occupied by the cockpit. The cockpit sidewalls were reinforced by four diagonal cross-braces.
Illustration from the parts list courtesy of ABC collection

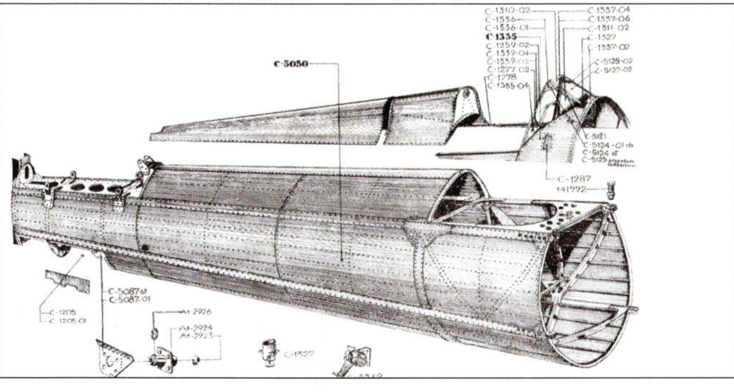

The rear semi-monocoque assembled with the spine, the cockpit sills and windscreen. The curved front shelf at the front of the cockpit served as a cross-brace and a support for the instrument panel. Illustration from the parts list courtesy of ABC collection

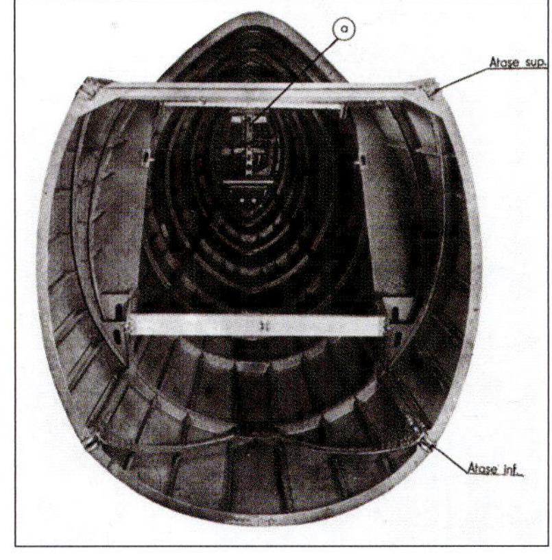

The rear fuselage viewed from the front. A = beam for attaching the oblique longeron of the tail fin to the rear fuselage; 'Ataşe sup' = Upper mounting points; 'Ataşe inf' = Lower mounting points. The two pins near the top of the diagonal shelves are the upper mounting points of the seat. The light-coloured horizontal beam in the lower half of the cockpit also served as a support for the floorboards. The rear pivot point for the aileron control bar is visible in the middle (slightly-off centre) of the front face of this beam.
Photo from the technical manual courtesy of ABC collection

Cockpit

Cockpit Floor and Sidewalls

The photos from the technical manual included here illustrate the cockpit of the prototype. The cockpit shown here remained generally the same on the production aircraft with a few minor changes.

This photo shows the lower part of the left cockpit sidewall. The lever at the top is the throttle lever. The black curved switch box behind the throttle is the magneto selector. The black dial is the oil temperature gauge. This was moved to the main instrument panel starting with No. 20. 'Pompa ambreiaj' = Pump clutch; 'Închis - Deschis' = Closed - Open; 'Tren ridicat' = Landing gear up; 'Tren afară' = Landing gear out; 'Siguranța pt. darea focului' = Gun firing safety; 'Voleți aripă' = Wing flaps; The knob of the landing gear selector was painted red and the knob of the flap selector was painted blue.
Photo from the technical manual courtesy of ABC Collection

This photo shows the forward lower part of the right cockpit sidewall. 'Extinctor comandat' = Manual fire extinguisher; 'Robinet de sig. tren' = Landing gear safety valve; 'Pompa de mână' = Manual pump (Hydraulic); 'Comanda scaunului' = Seat lever; 'Ridicarea scaunului' = To raise the chair; 'Scoborâre' = To lower the chair; 'Închis - Deschis' = Closed - Open; 'Voleți Naca' = Engine cooling gills; a = Chair position lever; b = Notched dial; c = Tooth for securing the lever a; d = Lever for releasing the tooth c; e = Pivoting axle for the chair position lever. The knob of the engine cooling flap control was painted blue.
Photo from the technical manual courtesy of ABC Collection

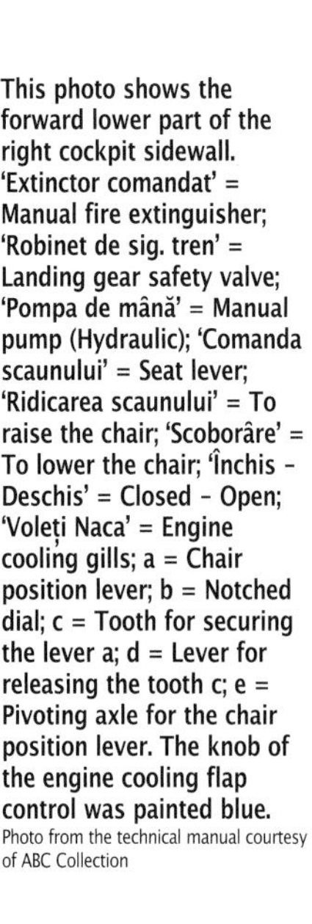

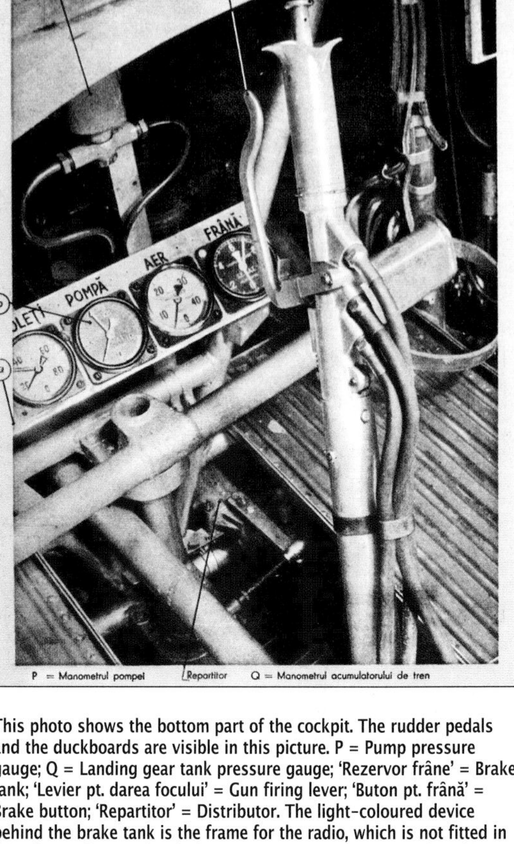

This photo shows the bottom part of the cockpit. The rudder pedals and the duckboards are visible in this picture. P = Pump pressure gauge; Q = Landing gear tank pressure gauge; 'Rezervor frâne' = Brake tank; 'Levier pt. darea focului' = Gun firing lever; 'Buton pt. frână' = Brake button; 'Repartitor' = Distributor. The light-coloured device behind the brake tank is the frame for the radio, which is not fitted in this photo. The panel with the pressure gauges is truncated in this picture as one instrument is missing from the left side.
Photo from the technical manual courtesy of ABC Collection

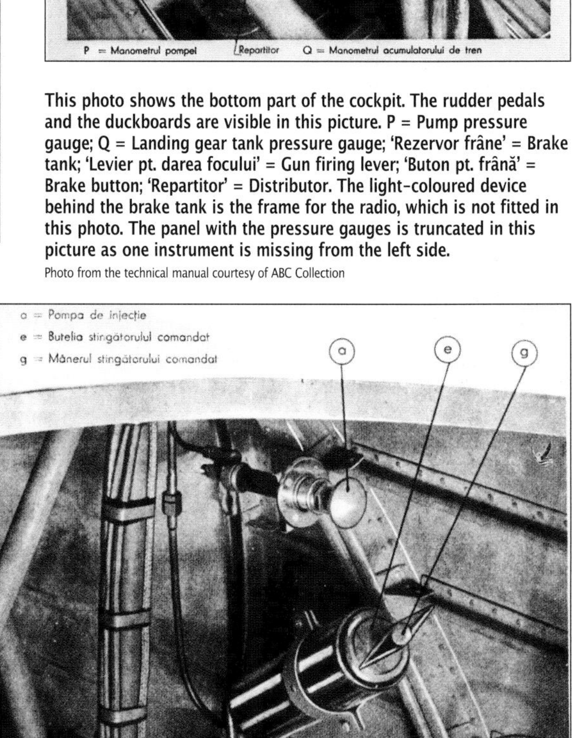

This photo shows the forward lower part of the right cockpit sidewall. a = Injection pump (choke); e = Manual fire extinguisher bottle; g = Handle of the manual fire extinguisher. On the I.A.R.80-C and I.A.R.81-C, the space above the fire extinguisher was used for mounting the cannon ammunition counters.
Photo from the technical manual courtesy of ABC Collection

I.A.R.80 Cockpit

These drawings depict an early I.A.R.80 series 1-20 cockpit, retro-fitted with the Dräger oxygen flow regulator. The cockpit of the later series was largely similar with few changes in the instrument panel and circuit breaker layouts. Please note that in the view showing the rear half of the cockpit, the headrest was omitted in order to show the structural elements with more clarity.

A. Right side of the cockpit viewed as if the left sidewall was removed;
B. Rear half of the cockpit viewed from the middle;
C. Front half of the cockpit viewed from the middle;
D. Left side of the cockpit viewed as if the right sidewall was removed;
E. Cockpit sills and windscreen viewed from above;
F. Cockpit viewed from above with the cockpit sills and windscreen removed.

1. Antenna mast
2. Rudder and brake steering assembly
3. Oil filter
4. Pneumatic distributor for the brakes
5. Manual fire extinguisher
6. Manual injection pump
7. Forward fuselage frame
8. Cockpit lamp
9. Canopy lock
10. Canopy slide bars
11. Main electrical switch
12. Headrest (early type pictured)
13. Lockable pulley for the seatbelt cable
14. Map case
15. Elevator trim tab adjustment wheel
16. Power regulator for the radio
17. Signal flare storage box
18. Landing gear safety valve
19. Machine gun trigger for all aircraft fitted with six pneumatically-fired 7.92 mm guns
20. Brake button
21. Lever for opening/closing the engine cooling gills. The knob of this lever was painted blue
22. Seat adjustment lever
23. Breathing mask storage box
24. Circuit breakers
25. Radio headset socket
26. Breathing mask heater (for aircraft No. 1 to 100)
27. Stowed manual starter crank for the engine
28. Control for the seatbelt cable
29. Klaxon, this gave a warning sound when the landing gear was retracted/deployed
30. Canopy handhold
31. Propeller pitch control
32. Oxygen socket for the breathing mask
33. Magneto selector switch
34. Valve for the hydraulic pump clutch
35. Pressure reducer
36. Pneumatic selector/safety valves for the machine guns on all aircraft fitted with six pneumatically-fired 7.92 mm guns
37. Duckboard frame
38. Throttle lever
39. DBU filter control lever - fuel mixture control
40. Dräger Type 10-20B oxygen flow regulator
41. Compressed air bottle for the brakes
42. Oxygen bottle
43. Compressed air bottle
44. Cockpit ventilation
45. Mechanical control for the landing gear lock
46. Landing gear hydraulic selector. The knob of this lever was painted red.
47. Landing flaps hydraulic selector. The knob of this lever was painted blue.
48. Late-type control 'spade grip' for all aircraft fitted with electrically-fired armament and retrofitted to all early-type aircraft re-armed with cannons
49. Manual hydraulic pump
50. Panel with pressure gauges (truncated to show the rudder and brake controls)
51. Panel with pressure gauges
52. Main instrument panel
53. Left instrument panel
54. Right instrument panel
55. Fuel cut-off levers
56. Ring and bead gunsight
57. Gunsight ring
58. Gunsight bead
59. Sandow for seat adjustment
60. Swing arm connecting the top seat bracket with the support/pivot point on the diagonal fuselage frame
61. Antenna wire
62. Rudder pedals
63. Rear view mirror
64. Lower seat bracket
65. Seat adjustment lever arm
66. Seat back rest. The oblong holes near the bottom are for the early-type shoulder straps
67. Seat pan with space for parachute pack
68. Diagonal seat brace
69. Elevator control rod
70. Elevator control bell crank
71. Rudder control cables
72. Aileron control rod
73. Seat support/pivot point on diagonal fuselage frame
74. Radio support frame
75. Diagonal fuselage frame
76. Flare pistol chute
77. Top cockpit shelf/brace

Based on drawings by Dan Iloiu, modified by Radu Brînzan ©2010

Control Panel

Prototype instrument panel. 1 = Fuel cut-off levers (mounted on the horizontal cockpit shelf); 2 = Position light switch; 3 = Fire warning; 4 = Air pump for fuel gauge

Instrument panel of the I.A.R.80 of the 1-20 series. This was identical to the prototype instrument panel, but the top of the central instrument panel was cut straight to improve the view of the ring and bead gunsight

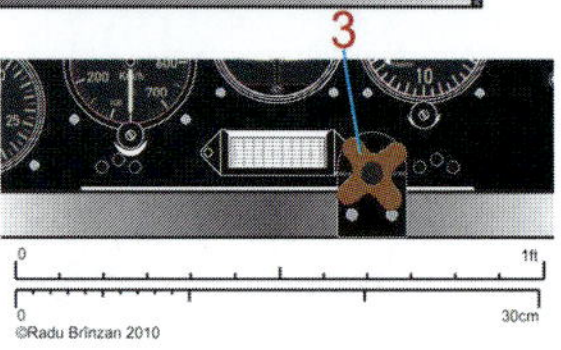

Standard instrument panel used on the I.A.R.80, I.A.R.80-A, I.A.R.80-B, I.A.R.80-C. The large round hole in the middle of the main instrument panel is for the gunsight cable. 1 = Mount for the optical gunsight; 2 = Compass correction chart; 3 = Control for the second oil radiator in the left wing introduced starting with I.A.R.80-C No. 251

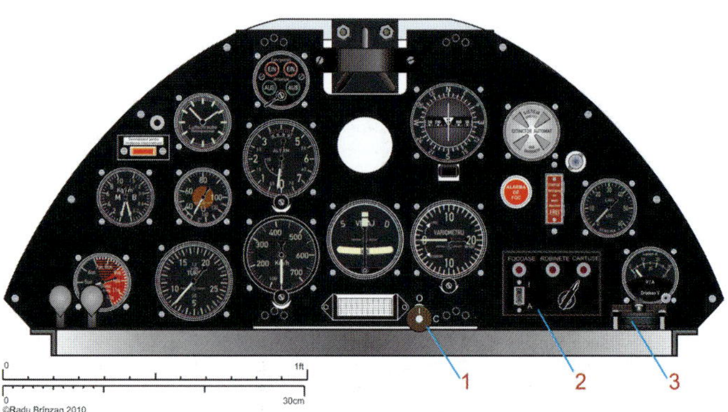

I.A.R.81 instrument panel. This was similar to the previous instrument panel but was equipped with bomb control box. 1 = Wing bomb carrier switch. By engaging this switch, the bombs could be launched at the same time as the main bomb. 2 = Bomb control box. 3 = Clock, moved to the horizontal cockpit shelf (not fitted to all aircraft). Photo courtesy of ABC Collection

Left side of the prototype instrument panel. a = Intake pressure gauge (boost); b = Siemens quadruple gauge (oil and fuel pressure gauge, input & output oil temperature); c = Tachometer; d = Switch for the position lamps on the wings and tail fin; e = Prop pitch indicator; f = Optical indicator for Pitot tube heater; g = Altimeter; h = Airspeed gauge; i = Compass; k = Illumination lamp for the instrument panel; l = Longitudinal inclinometer; m = Variometer; n = Pressure gauge of the automatic extinguisher; o = Fire warning; t = Optical landing gear indicator; q = Slip and bank indicator; z = Propeller pitch control Courtesy of ABC Collection

Cdr.Av. Gorelea and Cpt.Av. Ionică Teodorescu stand in front of an unidentified I.A.R.80 of the 1-20 series as indicated by the large inclinometer on the left side of the main instrument panel and the general arrangement of the instruments
Courtesy of ABC Collection

This photo of an I.A.R.-80 of Esc.47Vt./Gr.9Vt. provides a wealth of information about items on the left side of the instrument panel that are not usually visible in photos such as the cockpit light, the two fuel cut-off levers and the propeller pitch control. The aircraft is equipped with a Telereflex gunsight. Note the light colour of the cockpit, most likely the same as the underside colour. The ring under the windscreen is the canopy lock release. Courtesy of ABC Collection

The absence of the armoured windscreen and the bomb controls identify this aircraft as an I.A.R.81. The bomb selector panel is visible under the starter switch. The fire alarm cap is open and hanging by a chain over the bomb selector. This aircraft is equipped with a Goerz gunsight and 'spade grip' control stick. The presence of the control tap for the second oil cooler indicates that this is an aircraft of the 231-240 series or an earlier I.A.R.81 retrofitted with a second oil cooler
Courtesy of ABC Collection

Instrument panel of an I.A.R.81-C. This was identical to the I.A.R.81 instrument panel with the exception that the switch for the wing bomb carriers was removed (moved to the control stick) and a tap controlling the second oil cooler was mounted in its place

©Radu Brînzan 2010

This is a photo of an I.A.R.81-C as identified by the armoured windscreen. This aircraft is equipped with a Goerz gunsight. The ring on the left side of the canopy rail was the external canopy lock release. The internal canopy lock release toggle can be seen just to the right of the canopy rail, aligned with the ring. The fire alarm cap is open and hanging by a chain over the bomb selector. The control tap for the second oil cooler is partially visible
Courtesy of ABC Collection

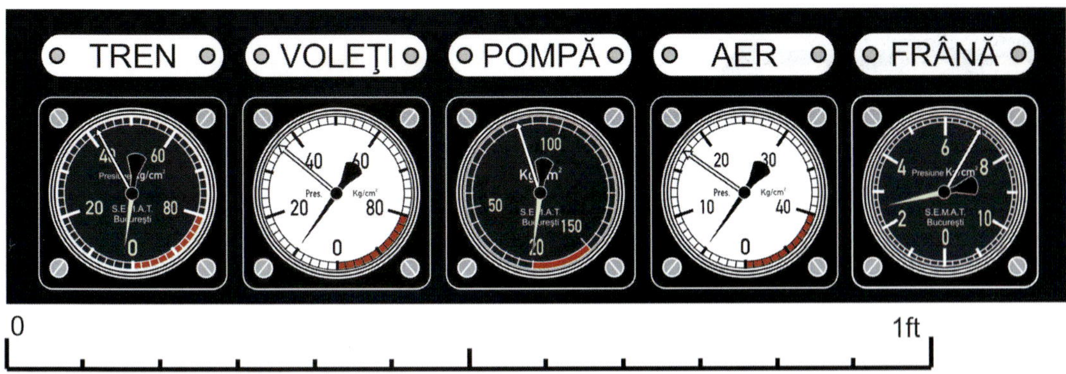

All aircraft of all series were equipped with a bank of pressure gauges that was placed above and behind the rudder pedal bar. These were labelled from left to right; 'Landing gear', 'Flaps', 'Pump', 'Air', 'Brakes'

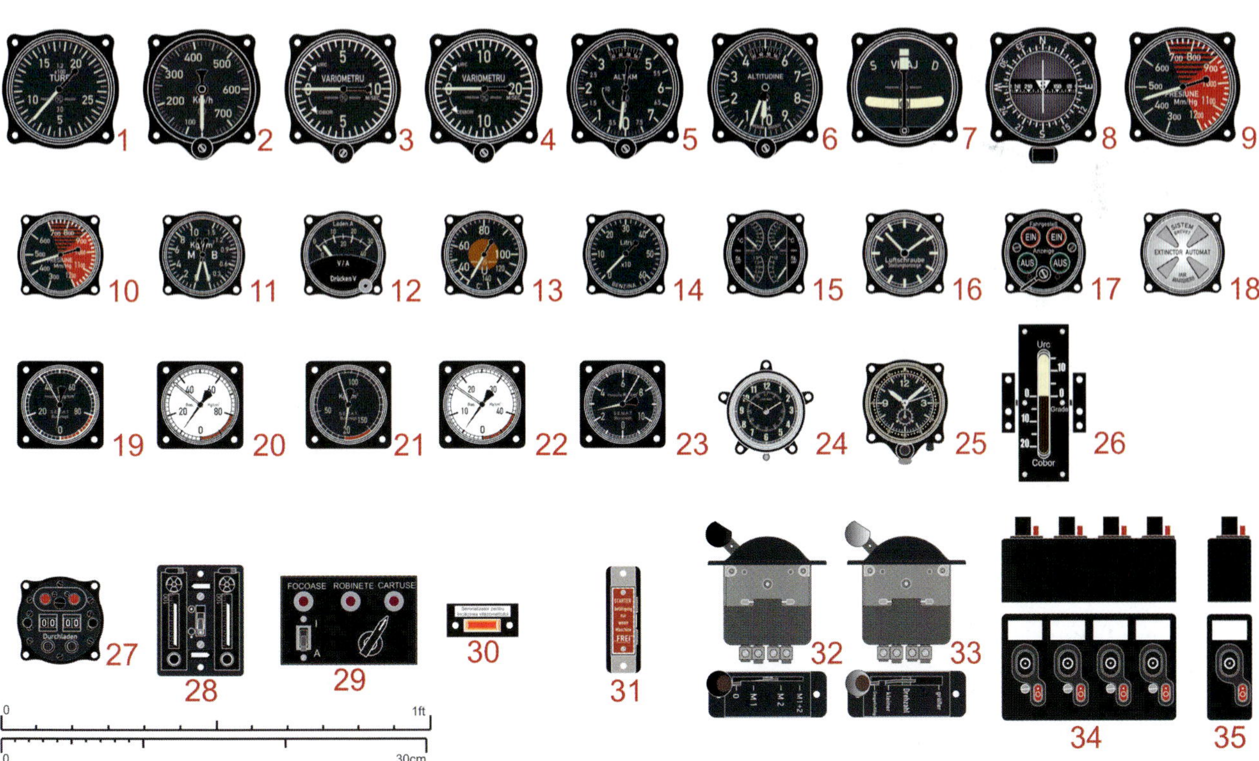

©Radu Brînzan 2010

Instruments used on the I.A.R. 80/81: 1 = Tachometer; 2 = Airspeed gauge; 3 = 10 m/sec variometer; 4 = 20 m/sec variometer; 5 = Coarse altimeter; 6 = Fine and coarse altimeter; 7 = Bank and slip indicator; 8 = Compass; 9 = Early boost gauge (large); 10 = Late boost gauge (small); 11 = Dual oil and fuel pressure gauge; 12 = Volt-Ampere meter [Pull toggle: Ampere / Push toggle: Volt]; 13 = Oil temperature; 14 = Fuel gauge; 15 = Quadruple gauge [oil & fuel pressure gauge / input & output oil temperature]; 16 = Propeller pitch control; 17 = Landing gear position indicator [Red = 'L/G up' / Green = 'L/G down' / Red & Green = 'L/G retracting/extending']; 18 = Fire extinguisher pressure gauge; 19 = Hydraulic pressure gauge for chassis retraction gear; 20 = Hydraulic pressure gauge for flaps; 21 = Pressure gauge for hydraulic pump; 22 = Compressed air pressure gauge; 23 = Brake air pressure gauge; 24 = Longines clock; 25 = Junghans clock with stopwatch; 26 = Linear inclinometer; 27 = ZVK-FF ammunition counter and arming control for Ikaria cannons [this was mounted on the left sidewall above the fire extinguisher]; 28 = SZKK2 ammunition counter and arming control for MG151/20 cannon [this was mounted on the left sidewall above the fire extinguisher]; 29 = Bomb control box ['Focoase' = Fuses, 'Robinete' = Valves, 'Cartuşe' = Bullets]; 30 = Pitot tube heater indicator lamp; 31 = Starter switch, underneath this flap was a handle that was pulled and turned to start the engine; 32 = Magneto switch; 33 = Propeller pitch control ['größer' = Coarse, 'drehzahl' = Revolutions count, 'kleiner' = Fine, 'segelstllg.' = Feathered] ; 34 = Bank of four circuit breakers; 35 = Single circuit breaker.

Gunsights

Ring and Bead Gunsight

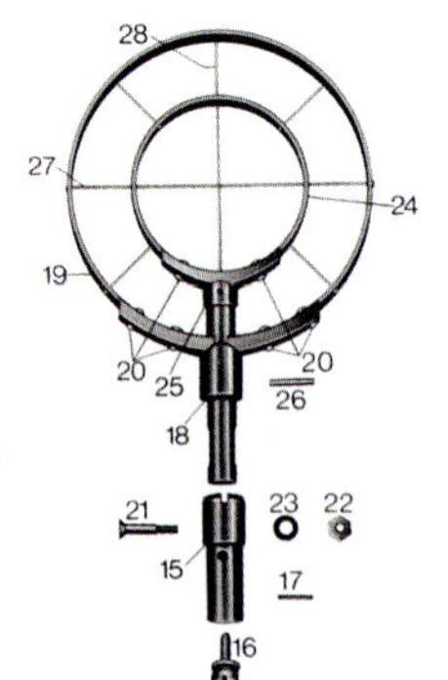

The 'ring' of the 'ring and bead' gunsight was the type fitted to Mauser flexible machine guns. The height of the ring was adjustable.
Courtesy of Edwin Wiedermer

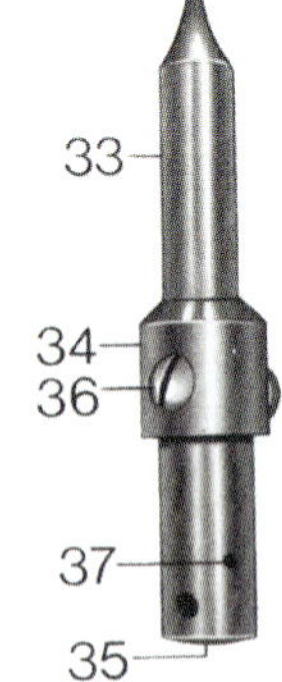

The height of the 'bead' of the 'ring and bead' gunsight was also adjustable.
Courtesy of Edwin Wiedermer

'Goerz and Telereflex Gunsights'

Initially, the I.A.R.80 was fitted with a basic 'ring and bead' sight, but starting with I.A.R.80 No.21, this was replaced by the Goerz GM2 gunsight. This gunsight was manufactured under licence by C.P. Goerz in Vienna and was based on the Barr & Stroud GM2 gunsight also known in the United Kingdom as 'Reflector Sight Mark II'. This was an optical gunsight that could be adjusted for accurate firing. The graticule projected on the reflector glass was a 'ring" bisected by a 'cross'. The horizontal line of the cross had a gap in the middle. Aiming was carried out by setting the two dials at the bottom of the gunsight. The top dial set the required firing range and the bottom ring set the wingspan of the target.

By making these adjustments, the gap in the horizontal line of the 'cross' was widened or narrowed and was thus set to the optimum firing setting. When the wingspan of the target filled this gap, the guns could be fired for the optimum result. The 'ring' of the graticule indicated the amount of deflection allowed for hitting a target crossing it. As the supplies of Goerz gunsights were becoming increasingly unreliable, the factory looked for an alternative sighting device. The Telereflex gunsight was developed by PREROM and I.O.R. in Romania based on the Goerz design. In fact, the two gunsights were identical internally and operated in the same manner. The main difference between the Goerz GM2 and Telereflex gunsights was in the mounts for the reflector glass and the sun visor.

'Goerz GM2 Gunsight fitted to an I.A.R.81-C as identified by the armoured windscreen'
Photo courtesy of ABC Collection

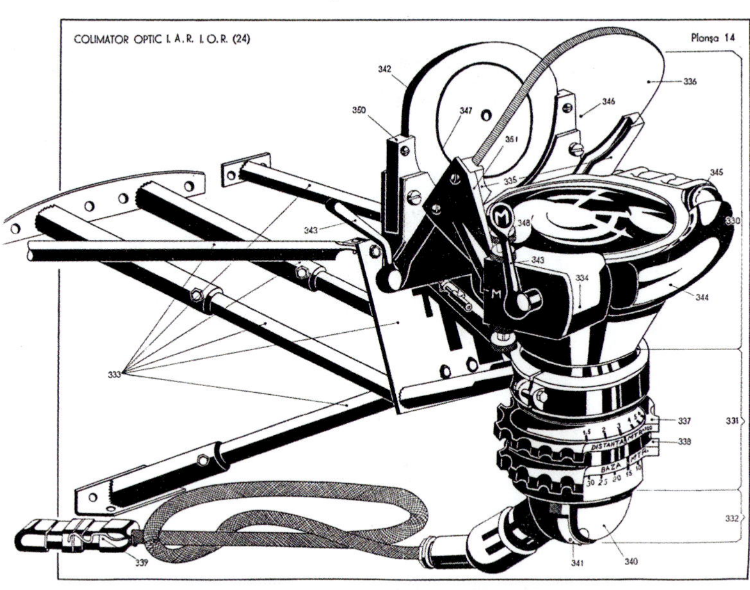

The Telereflex gunsight, the mounting bracket and the power cable. Drawing from the technical manual courtesy of Dan Melinte

A Telereflex gunsight is on display in the aviation exhibition of the Military Museum in Bucharest. The reflector and sun visor glass plates are missing from this sight.
Photo by Radu Brînzan

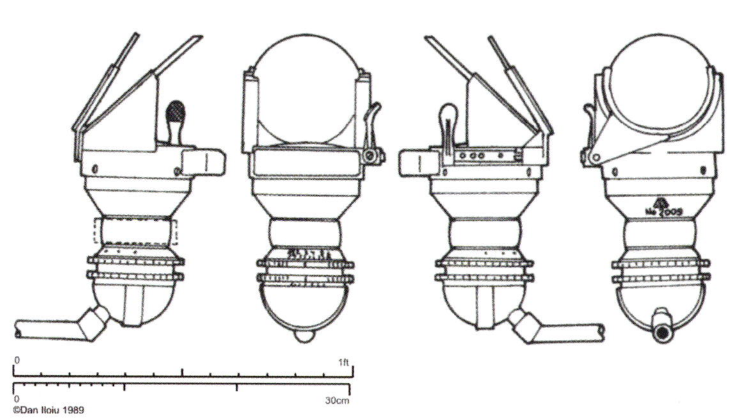

Goerz GM2 Gunsight
Drawing by Dan Iloiu

Headrests

Headrests and head armour fitted to the early I.A.R.80 and 81 are a veritable minefield. There were two types of head armour plate and three types of head cushion. Although it is possible to pinpoint the manufacturing batch when a change of head cushion or head armour was implemented, one must keep in mind that as new types were introduced, they were also retrofitted to earlier types in a variety of combinations, most likely due to availability of spares and parts. The headrests were standardised starting with No. 251, but in the case of aircraft up to No. 250, one needs to consult the existing photographs to ascertain the exact type of head armour and head cushion fitted. The head armour plate was painted dark green. The colour of the cushion varied depending on the material used and the amount of wear-and-tear. The

Lt.Av. Nicolae Linburg smiles from the the cockpit of I.A.R.80-B No. 225. This clearly shows the rubber seal and external frame typical for modified mid-production headrest introduced from No. 181. Note the mid-production type of cushion with a 'flat' top. This aircraft is fitted with the early-type seatbelts.
Photo courtesy of ABC Collection

cushion was mostly leather-covered, but some photographs seem to suggest that at least some cushions were covered with khaki-coloured fabric.

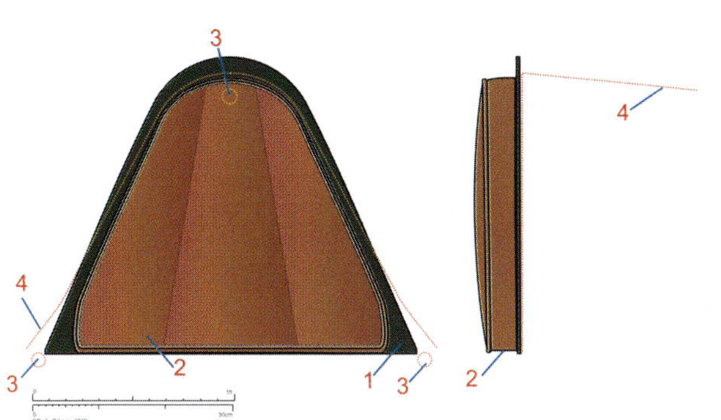

Early-production headrest. This was the type used on all aircraft up to No. 130. 1 = Head armour plate, perpendicular to the datum line; 2 = Headrest cushion; 3 = Canopy rail position; 4 = Outline of the fuselage spine behind the armour plate

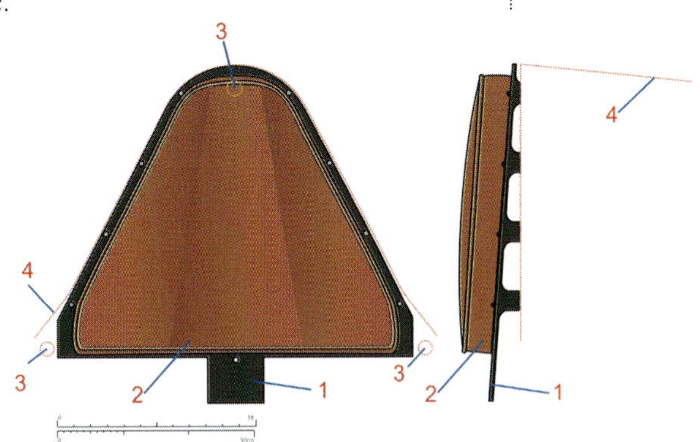

Mid-production headrest. This was introduced beginning with No. 131. The head armour was placed at an angle matching the angle of the seat back rest and thus making it more comfortable for the pilot. The top edge of the cushion was relatively 'flat'. 1 = Head armour. The rectangular protrusion at the bottom of the head armour covered the pulley of the seatbelt tensioning cable, which passed through the perforation in the armour plate; 2 = Headrest cushion; 3 = Canopy rail position; 4 = Outline of the fuselage spine behind the armour plate

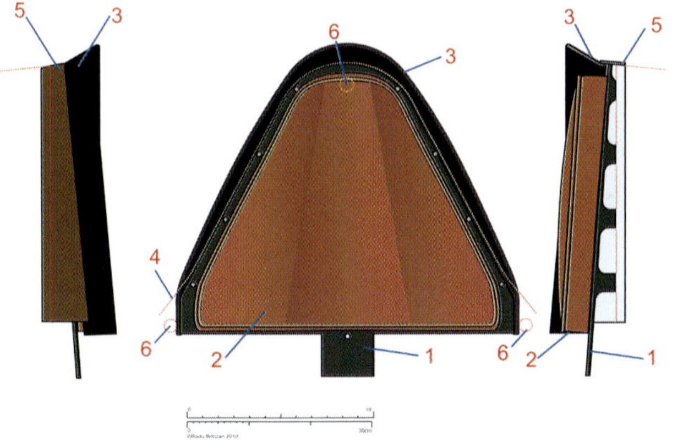

Modified mid-production headrest. This was introduced beginning with No. 181. The head armour and head cushion were the same as the type introduced with No. 131 but the headrest was fitted with a rubber seal that provided a certain amount of draft protection. 1 = Head armour; 2 = Headrest cushion; 3 = Rubber seal; 4 = Outline of the fuselage spine behind the armour plate; 5 = External headrest frame; 6 = Canopy rail position

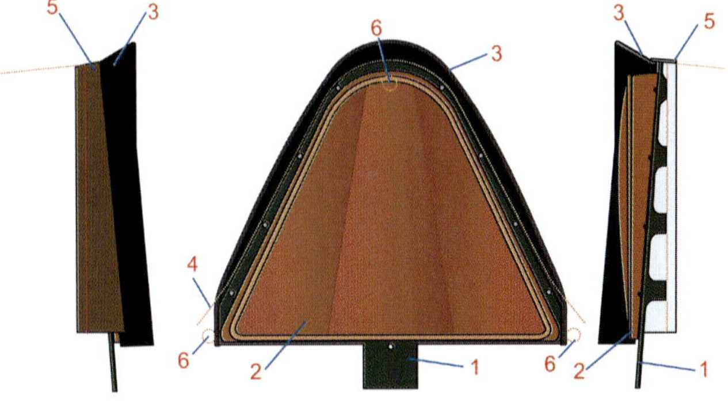

Late-production headrest. This was introduced beginning with No. 251. The head armour was the same as the type introduced with No. 131. The head cushion was modified with a rounder top and a wedge-shaped side profile. The headrest was also fitted with a rubber seal that provided a certain amount of draft protection. 1 = Head armour; 2 = Headrest cushion; 3 = Rubber seal; 4 = Outline of the fuselage spine behind the armour plate; 5 = External headrest frame; 6 = Canopy rail position

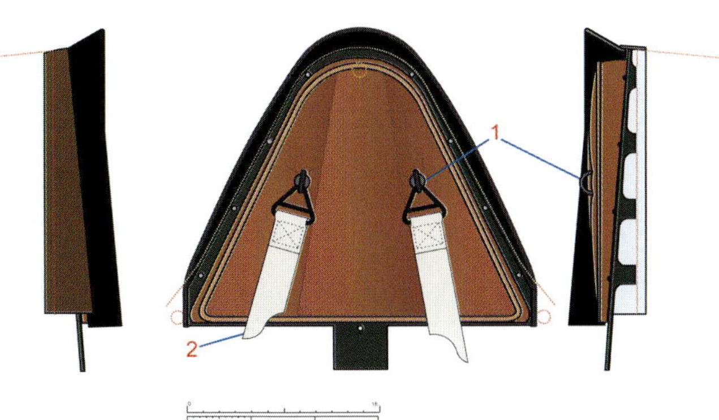

Final-production headrest. This is identical to the type of headrest introduced with No. 251, but equipped with seatbelt fittings. The late-type seatbelts were secured to loops bolted through the head cushion and head armour. This type of headrest was introduced on the production line and retrofitted to most aircraft from June 1943. 1 = Seatbelt attachment loops; 2 = Seatbelts

Slt.Av. Ioan Mihăilescu sits in the cockpit of an unknown I.A.R. Note the angled mid-production head armour plate and the mounting "tabs". However, it must be pointed out that this is fitted with a late-production cushion, as indicated by the rounder top and the wedge-shaped side profile. This is the only known photograph of this type of head armour in this configuration. This photograph is unique because all the aircraft initially fitted with this type of headrest were retrofitted with the rubber seal and frame associated with the late-production headrests. Courtesy of ABC Collection

"

Canopy

The canopy used on the I.A.R.80 was a single piece of blown clear material supported by a tubular frame around the edges. The clear material is identified in the technical manual as 'Plexiglas'. This type of canopy was used on all series of I.A.R.80/81. The canopy was manually operated and slid on three tubular rails, one on each cockpit sill and one behind the head armour under the spine. This arrangement presented a number of safety issues. If any of the rails were damaged or bent in combat or in an accident, the canopy could no longer slide freely. At speeds exceeding 250 km/h, the canopy was squeezed by the airflow and could no longer open. Starting with No.241 an emergency release pneumatic ram was fitted at the back of the canopy inside the spine to assist with sliding back the canopy at high speeds. This emergency device was also retrofitted to many existing airframes. This system was not regarded as a solution by the pilots because the system could fail in the case of loss of pneumatic pressure. Pilots continued to demand a canopy release system that jettisoned the entire canopy assembly including the rails, similar to the system used on the Bf.109, but such a system was never introduced before the end of the operational use of the aircraft.

Lt.Av. Nicolae Limburg sits in the cockpit of an I.A.R.81-C. This shows how the rear of the canopy curves inwards towards the back to conform to the fuselage spine.
Photo courtesy of ABC Collection

Adj.Av. Vasile Chiriac smiles from the cockpit of an I.A.R.81-C. This photo shows the canopy rails and the sliding ring at the bottom of the forward canopy frame.
Photo courtesy of ABC Collection

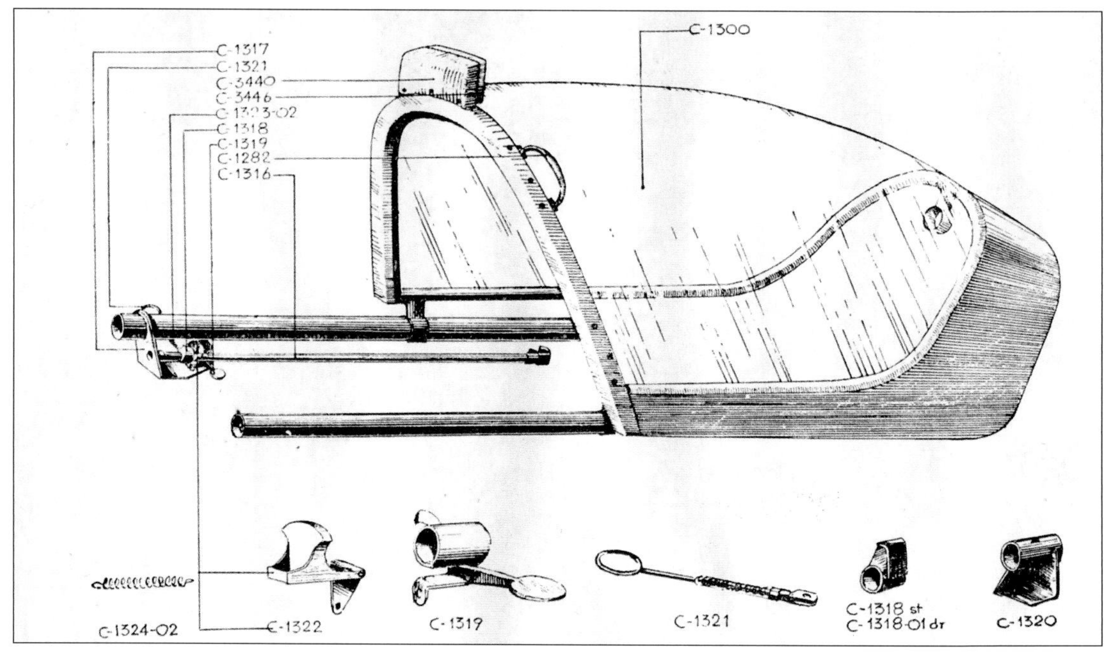

The canopy components and the two side rails. C-1319 is the internal canopy latch. C-1321 is the external canopy lock toggle.
Illustration from the technical manual courtesy of ABC Collection

Windscreen

The windscreen went through a number of changes during the service life of the I.A.R.80/81 but externally it remained the same. Whereas the first type of windscreen was made of Plexiglas, a reinforced windscreen made of layered glass was introduced beginning with I.A.R.80-A No. 131, as illustrated in this photo. The purpose of the 'kink' at the bottom of the windscreen side panels was to allow the fuselage to bulge slightly to clear the lateral instrument panels. Note that the glass of the side panels is split in that area. Photo courtesy of Jose Fernandez

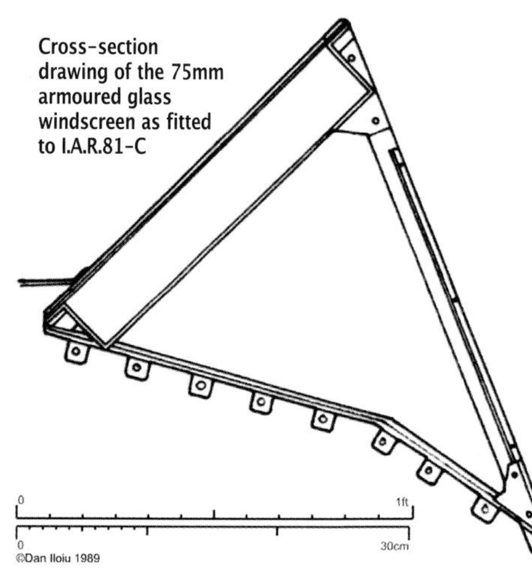

Cross-section drawing of the 75mm armoured glass windscreen as fitted to I.A.R.81-C

Wings

Four-Gun Wing

The first type of wing used on the I.A.R.80 was armed with four 7.92 mm FN Browning machine guns. The wing was based on the very popular NACA-32012 profile. The wing had 36 ribs, 18 ribs in each half-wing. Ribs 2, 6, 9, 11, 14 and 16 [counting from the wing tip] that held the control surface hinges and any other load-bearing structures were reinforced with double-T sections. Because the landing gear retracted inside the wing, in order to preserve the structural strength of the wing, the wing centre was provided with a cross-shaped box-type beam and with double-T frames between ribs 12 and 17. The wing spars were made of Dural. The section of the spar between ribs 1 and 11 had a double-T cross-section and the remaining section was box-shaped. The front wing spar was straight and the rear wing spar angled forward from the fuselage towards the wingtips. The angle of attack of the wings was of 2° relative to the aircraft datum line.

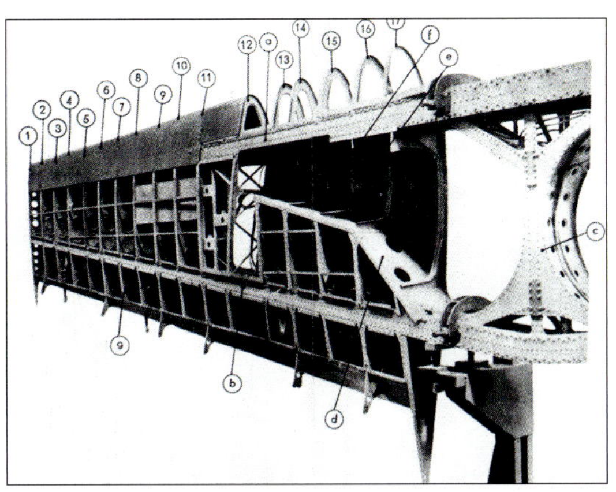

This photo shows the internal structure of the four-gun wing.
a = Front spar; b = Rear spar; c = Cross-shaped box-type beam;
d = Diagonal; e = Cut-outs in the ribs; f = L-shaped longitudinal strips to reinforce the skin; g = Wing rib; 1-17 = wing ribs.
Photo from the technical manual courtesy of ABC Collection

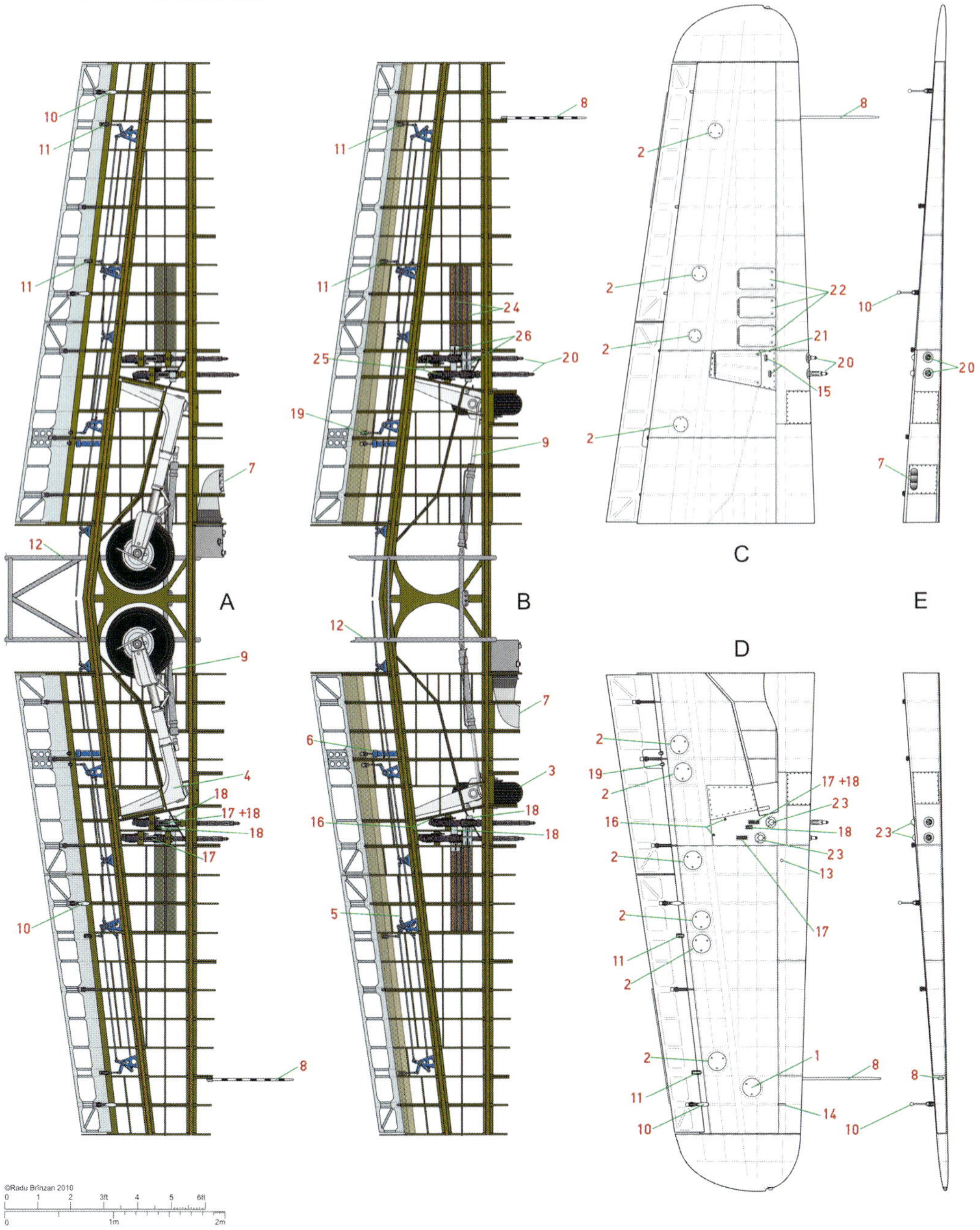

Four-Gun Wing: 4 x 7.92 mm FN Browning Machine Guns
Used on I.A.R. 80 Serial Numbers 1-50

A - Internal structure of the wing viewed from below;
B - Internal structure of the wing viewed from above;
C - Port wing viewed from above;
D - Port wing viewed from below;
E - Wing viewed from the front;
1. Service panel for Pitot Tube (port underside only);
2. Service panels for control linkage levers and couplings;
3. Landing gear extended;
4. Landing gear retracted – early "long leg" landing gear;
5. Control linkages;
6. Hydraulic ram for the flap;
7. Oil radiator intake;
8. Pitot Tube (port side only);
9. Hydraulic ram for the landing gear;
10. Aileron mass-balancing counterweights;
11. Aileron control linkages;
12. Internal fuselage frame (shown partially);
13. Hoist point;
14. Mooring ring;
15. Cooling vent for the gun barrel;
16. Manual gun charging toggle, 7.92mm machine gun;
17. Spent Ammunition chute, 7.92mm machine gun;
18. Spent belt link chute;
19. Flap-to-aileron linkage rod;
20. 7.92mm FN Browning machine gun;
21. Access hatch for machine guns;
22. Access hatch for ammunition;
23. Domed covers for the machine gun mounting and alignment fittings;
24. 7.92 mm machine gun ammunition;
25. Ammunition feed chute;

©Radu Brinzan 2010

Six-Gun Wing

The wing equipped with six 7.92 mm FN Browning machine guns was used by the I.A.R.80-A and I.A.R.81. Internally, the wing was almost identical to the four-gun wing but was strengthened with extra span-wise stringers.

On 2 May 1941, I.A.R. Braşov submitted a memo regarding the strength of the I.A.R.81 wing:

'The safety coefficient was 2, respectively higher than the 1.8 required by the German regulations or the 1.5 required by Handley Page. The calculation of the wing strength was carried out by taking into consideration the extra load that the wing had to absorb because of the dive brake. The maximum load at the root joint was calculated to 4,525 kg, compared to the 6,345 kg that the wing was subjected to during the static test. This resulted in a safety coefficient higher than 2.8 in the case of an aircraft equipped for dive bombing with a total weight of 2,860 kg at a speed of 660 km/h.'

Due to the change in the length of the landing gear demanded by the bomb equipment, there was a slight change in the angle of the diagonal brace behind the wheel well.

Early Six-Gun Wing: 6 x 7.92 mm FN Browning Machine Guns
Used on I.A.R. 80-A Serial Numbers 51-75

A - Internal structure of the wing viewed from below;
B - Internal structure of the wing viewed from above;
C - Port wing viewed from above;
D - Port wing viewed from below;
E - Wing viewed from the front;
1. Service panel for Pitot Tube (port underside only);
2. Service panels for control linkage levers and couplings;
3. Landing gear extended;
4. Landing gear retracted - early "long leg" landing gear;
5. Control linkages;
6. Hydraulic ram for the flap;
7. Oil radiator intake;
8. Pitot Tube (port side only);
9. Hydraulic ram for the landing gear;
10. Aileron mass-balancing counterweights;
11. Aileron control linkages;
12. Internal fuselage frame (shown partially);
13. Hoist point;
14. Mooring ring;
15. Cooling vent for the gun barrel;
16. Manual gun charging toggle;
17. Spent ammunition chute;
18. Spent belt link chute;
19. Flap-to-aileron linkage rod;
20. 7.92mm FN Browning machine gun;
21. Access hatch for machine guns;
22. Access hatch for ammunition;
23. Domed covers for the machine gun mounting and alignment fittings;
24. 7.92 mm machine gun ammunition;
25. Pneumatic firing mechanism;
26. Ammunition feed chute.

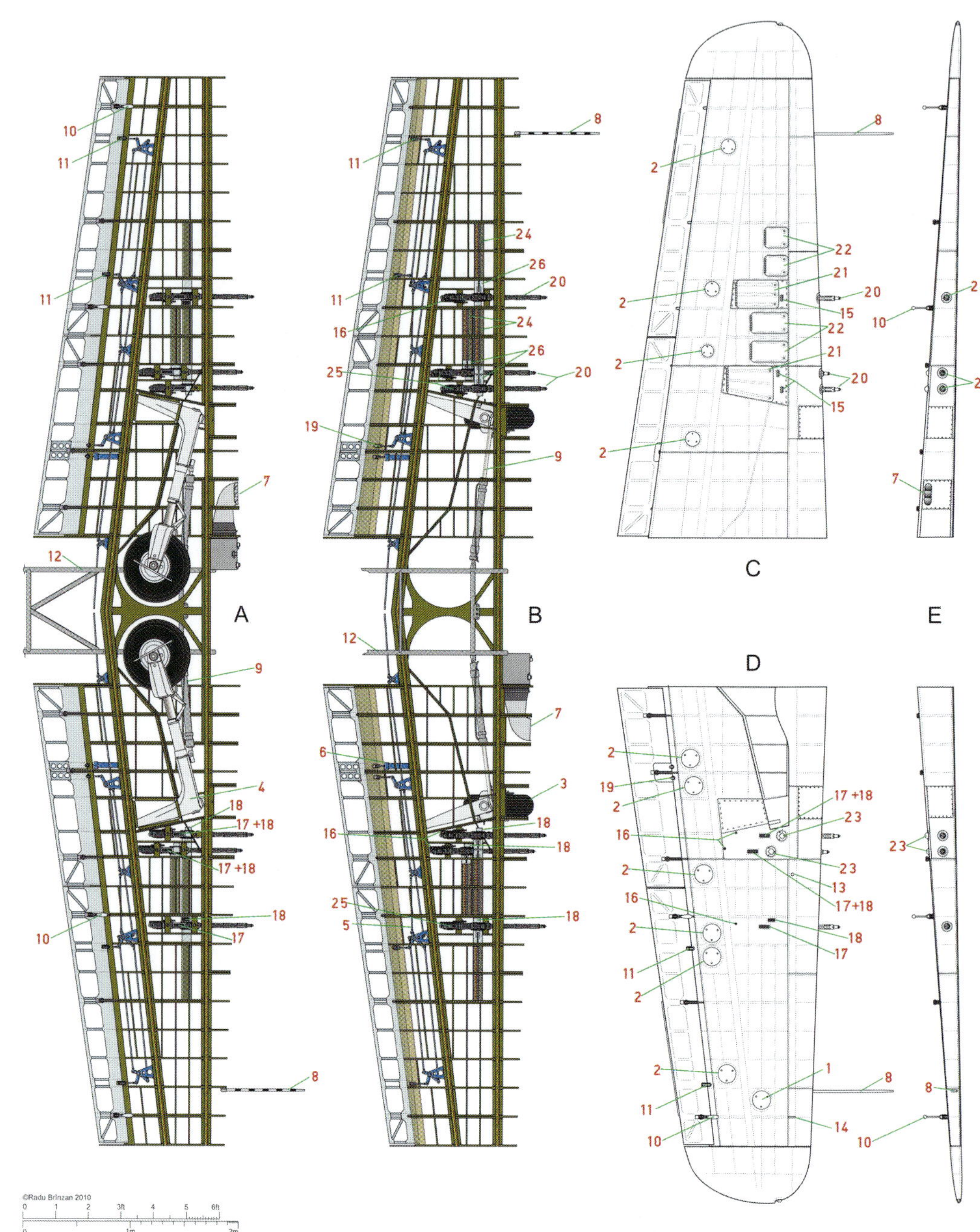

Late-type Six-Gun Wing

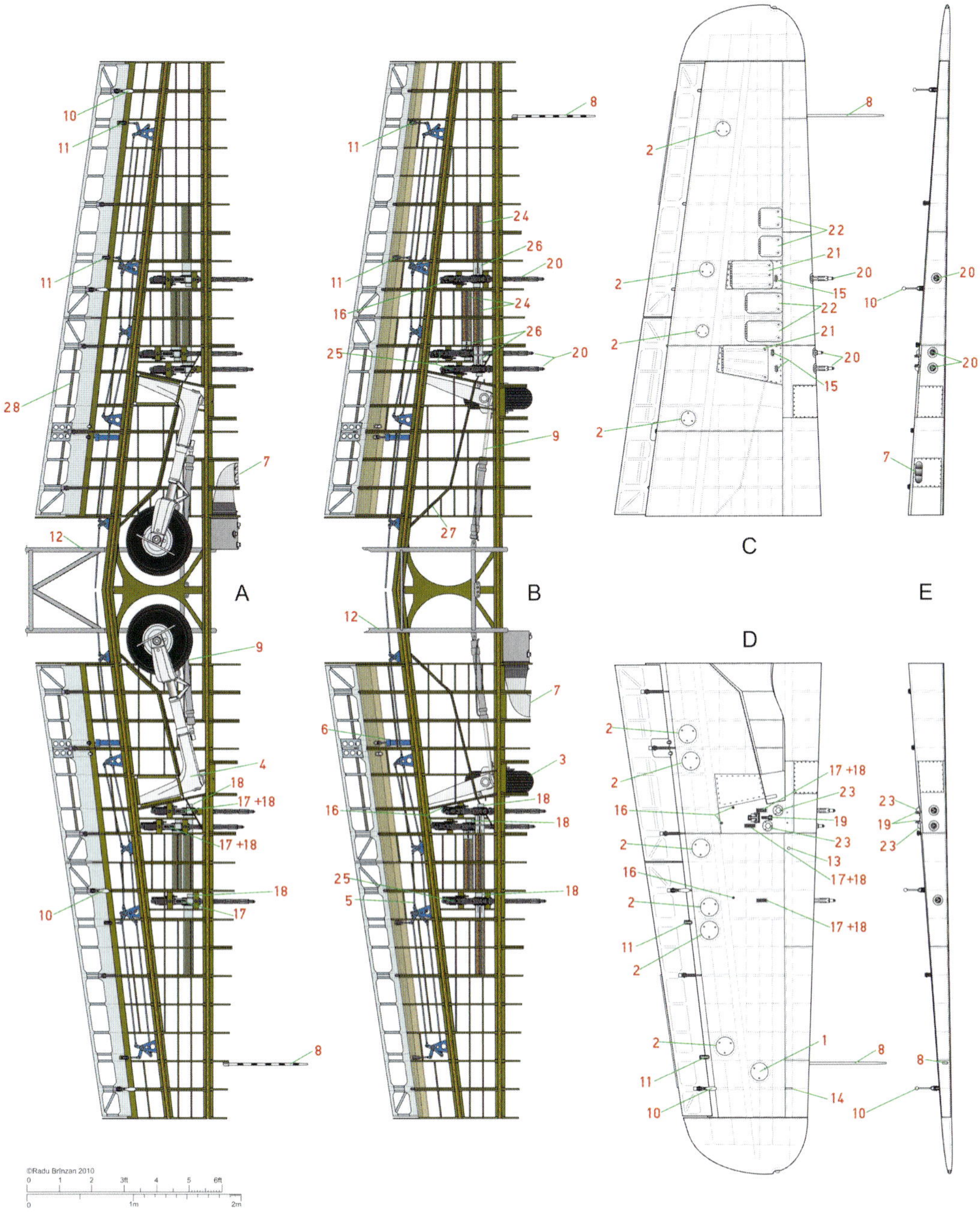

Late Six-Gun Wing: 6 x 7.92 mm FN Browning Machine Guns
Used on I.A.R. 81 Serial Numbers 91-105, 151-175 and 231-240
Also used without bomb racks on I.A.R. 80-A Serial Numbers 76-90, 106-150 and 176-180

A - Internal structure of the wing viewed from below;
B - Internal structure of the wing viewed from above;
C - Port wing viewed from above;
D - Port wing viewed from below:
E - Wing viewed from the front;
1. Service panel for Pitot Tube (port underside only);
2. Service panels for control linkage levers and couplings;
3. Landing gear extended;
4. Landing gear retracted - final "short leg" landing gear';
5. Control linkages;
6. Hydraulic ram for the flap;
7. Oil radiator intake;
8. Pitot Tube (port side only);
9. Hydraulic ran for the landing gear;
10. Aileron mass-balancing counterweights;
11. Aileron control linkages;
12. Internal fuselage frame (shown partially);
13. Hoist point;
14. Mooring ring;
15. Cooling vent for the gun barrel;
16. Manual gun charging toggle;
17. Spent ammunition chute;
18. Spent belt link chute;
19. Bomb racks. Only on I.A.R. 81;
20. 7.92 mm FN Browning machine gun;
21. Access hatch for machine guns;
22. Access hatch for ammunition;
23. Domed covers for the machine gun mounting and alignment fittings;
24. 7.92 mm machine gun ammunition;
25. Pneumatic firing mechanism;
26. Ammunition feed chute;
27. Angle of the inboard diagonal was changed in order to accommodate the wheel of the shorter landing gear leg. This feature was applied to all subsequent I.A.R. 80 and I.A.R. 81 aircraft beginning with No. 76;
28. Flap reinforcement bar. Fitted to all subsequent aircraft.

©Radu Brînzan 2010

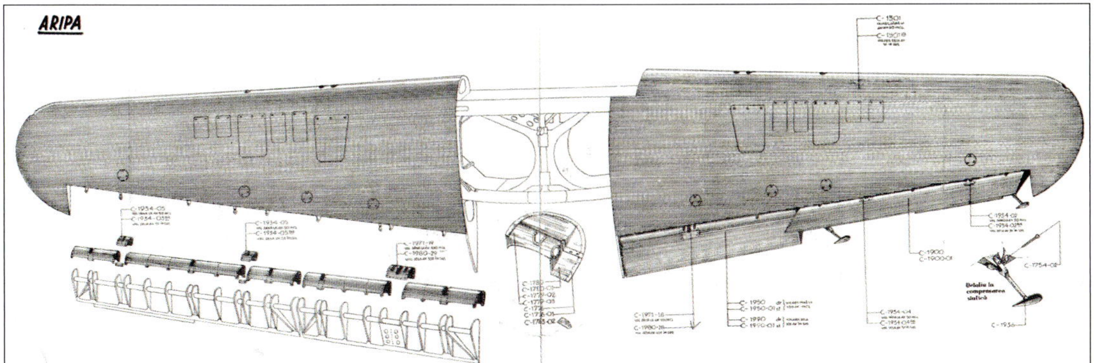

The internal structure of the six-gun wing and the skin
Illustration from the technical manual courtesy of ABC Collection

A – Internal structure of the wing viewed from below;
B – Internal structure of the wing viewed from above;
C – Port wing viewed from above;
D – Port wing viewed from below;
E – Wing used on No. 181 to 230 viewed from the front;
F – Wing used on No. 291 to 300 viewed from the front;
G. Scrap view of the second oil radiator used on No. 291 to 300;
1. Service panel for Pitot Tube (port underside only);
2. Service panels for control linkage levers and couplings;
3. Landing gear extended;
4. Landing gear retracted – final "short leg" landing gear;
5. Control linkages;
6. Hydraulic ram for the flap;
7. Oil radiator intake;
8. Pitot Tube (port side only);
9. Hydraulic ram for the landing gear;
10. Aileron mass-balancing counterweights;
11. Aileron control linkages;
12. Internal fuselage frame (partial view);
13. Hoist point;
14. Mooring ring;
15. Cooling vent for the gun barrel;
16. Manual gun charging toggle;
17. Spent ammunition chute;
18. Spent belt link chute;
19. 13.2 mm FN Browning machine gun;
20. 7.92 mm FN Browning machine gun;
21. Access hatch for machine guns;
22. Access hatch for ammunition;
23. Domed covers for the machine gun mounting and alignment fittings;
24. 7.92 mm machine gun ammunition;
25. Electrical gun-firing mechanism;
26. Ammunition feed chute;
27. Domed fairing for the 7.92 mm machine gun firing solenoid;
28. Access panel for the 31.2 mm machine gun firing mechanism extending behind the rear wing spar;
29. Access panel for arming the 13.2mm machine gun;
30. 13.2 mm machine gun ammunition;
31. Drop tank rack - not used on No. 181-200;
32. Cross-brace (removable for gun servicing).

Heavy Gun Wing

The wing equipped with four 7.92 mm FN Browning machine guns and two 13.2 mm FN Browning machine guns was fitted to the I.A.R.80-B. The first series of I.A.R.80-B had the same wingspan as the I.A.R.89-A, but starting with No. 212, the wingspan was extended to 11 metres increasing the lift area to 16.5 square metres. The spars were extended, the wing to fuselage attachment bolts were strengthened by 20% and the diagonal between ribs 4 and 5 was increased to 2.5 mm. The wing skin was increased from 0.8 mm to 1 mm.

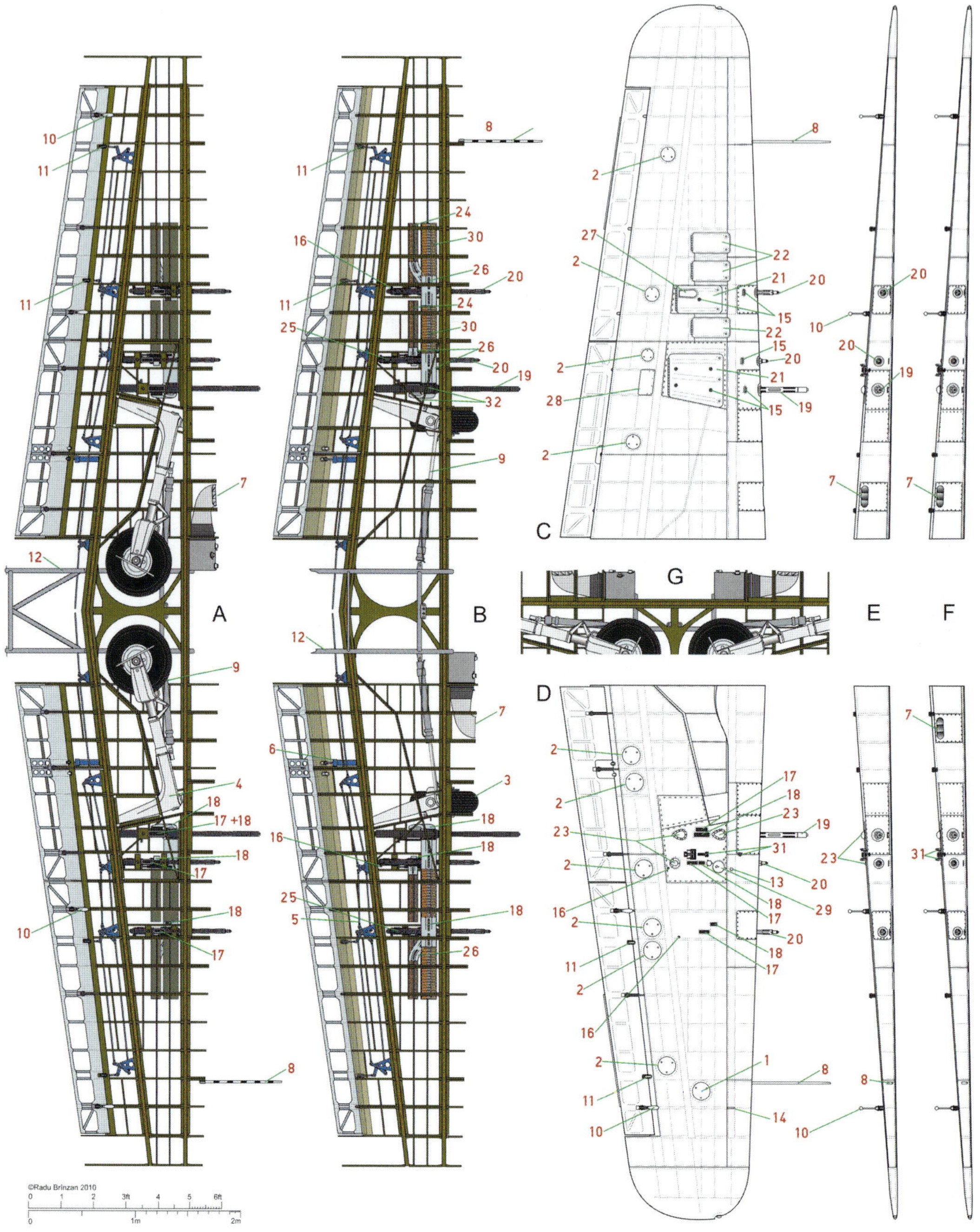

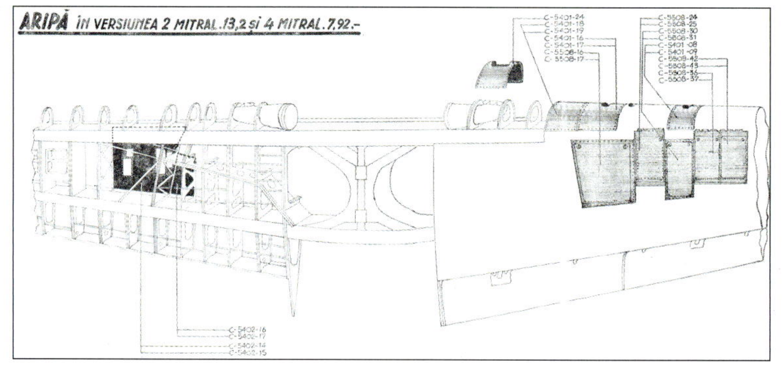

The internal structure and the panels used on the heavy-gun wing. It is interesting to note that this wing is shown with two oil coolers, which was used only on the last series of I.A.R.80-B
Photo courtesy of ABC Collection

©Radu Brînzan 2010

Ikaria Cannon Wing

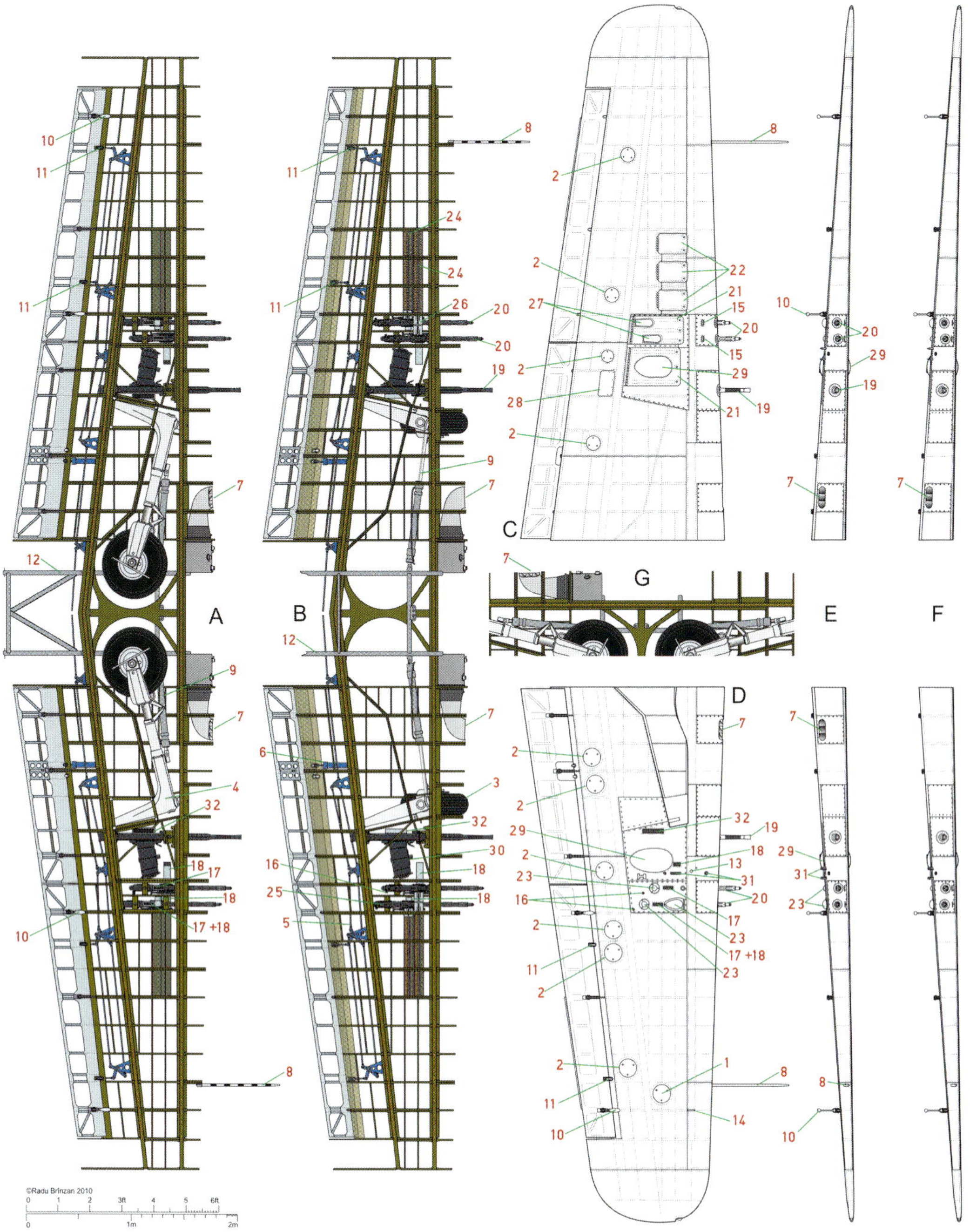

Ikaria Cannon Wing: 4 x 7.92 mm FN Browning Machine Guns & 2 x 20 mm Ikaria MG-FF Cannons
Used on I.A.R. 80-C Serial numbers 242-290. The I.A.R. 80-C Serial Numbers 242-250 were equipped with a single oil radiator.

A - Internal structure of the wing viewed from below;
B - Internal structure of the wing viewed from above;
C - Port wing used on the No. 251 to 290 viewed from above;
D - Port wing used on No. 251 to 290 viewed from below;
E - Wing used on No. 251 to 290 viewed from the front;
F - Wing used on No. 241 - 250 viewed from the front;
G - Scrap view of the single oil radiator used on No. 241 to 250;

1. Service Panel for Pitot tube (port underside only);
2. Service panels for control linkage levers and couplings;
3. Landing gear extended;
4. Landing gear retracted - final "short leg" landing gear;
5. Control linkages;
6. Hydraulic ram for the flap;
7. Oil radiator intake;
8. Pitot Tube (port side only);
9. Hydraulic ram for the landing gear;
10. Aileron mass-balancing counterweights;
11. Aileron control linkages;
12. Internal fuselage frame (shown partially);
13. Hoist point;
14. Mooring point;
15. Cooling vent for the gun barrel;
16. Manual gun charging toggle;
17. Spent 7.92 mm ammunition chute;
18. Spent 7.92 mm belt link chute;
19. 20 mm Ikaria MG-FF Cannon;
20. 7.92 mm FN Browning Machine gun;
21. Access hatch for weapons;
22. Access hatch for 7.92 mm ammunition;
23. Domed covers for the machine gun mounting and alignment fittings;
24. 7.92 mm machine gun ammunition;
25. Electrical machine gun-firing mechanism;
26. Ammunition feed chute;
27. Domed fairing for the 7.92 mm machine gun firing solenoid;
28. Access panel for the 20 mm cannon firing fittings extending behind the rear wing spar;
29. Domed fairing for the cannon ammunition drum;
30. 20 mm cannon ammunition drum;
31. Drop tank rack;
32. Chute for spent cannon rounds.

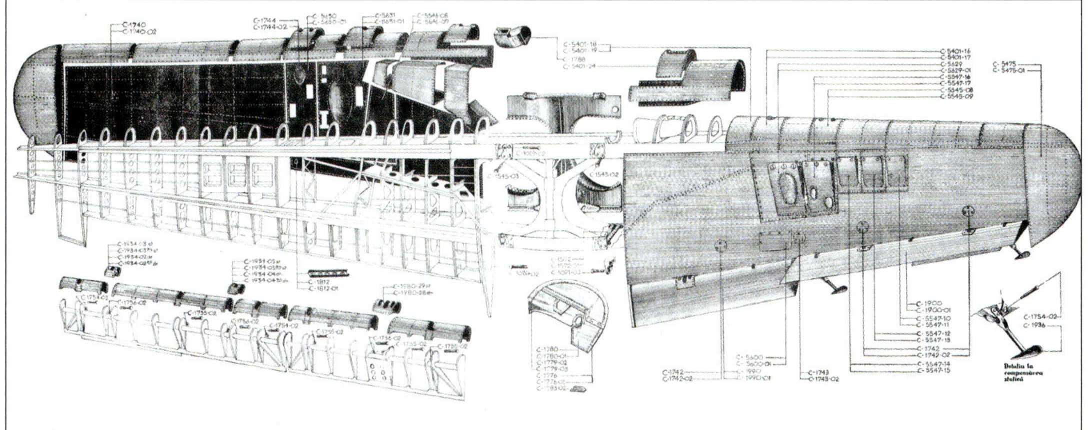

The components of the Ikaria cannon wing
Illustration from the technical manual courtesy of ABC Collection

Mauser Cannon Wing

The Mauser MG151/20 was fitted to the I.A.R.81-C, which was initially designed as a dive-bomber. However, the weight increase caused by adding the cannons led to doubts in relation to this role

Letter B-328/02.06.1943 regarding the possibility of arming the I.A.R.81 (No. 301-450) aircraft with Mauser cannons issued by I.A.R.-Braşov:

'The total weight increases to 3,000 kg. For this category, the aircraft is subjected to a supplementary load factor of n = 13, with static tests even indicating a load factor of n = 15. With one 225 kg bomb and two 50 kg bombs, the weight increases to 3,325 kg with a load factor of n=11.5, sufficient for bombers where the required load factor is n = 11. However, as the total load is very high ($3,325 \div 16.5 = 202$ kg/m²), the pilots flying the aircraft thus equipped must be chosen from among the best. Also because of this high load, the planes can only be used in the following roles: pure fighter, long-range (with drop tanks) escort for bombers or dive bombers with a maximum bomb load of 325 kg'.

The covers are being removed from I.A.R.81-C No. 323 of Esc.61Vt./Gr.6Vt on Popesti-Leordeni airfield in January 1944. This photo shows the bulges and spent ammunition/links chutes associated with the Mauser cannon wing. The wing bomb/drop tank carrier was located outboard of the 7.92 mm machine gun. The bottom landing gear covers were removed to prevent clogging with mud and snow.
Photo courtesy of ABC Collection

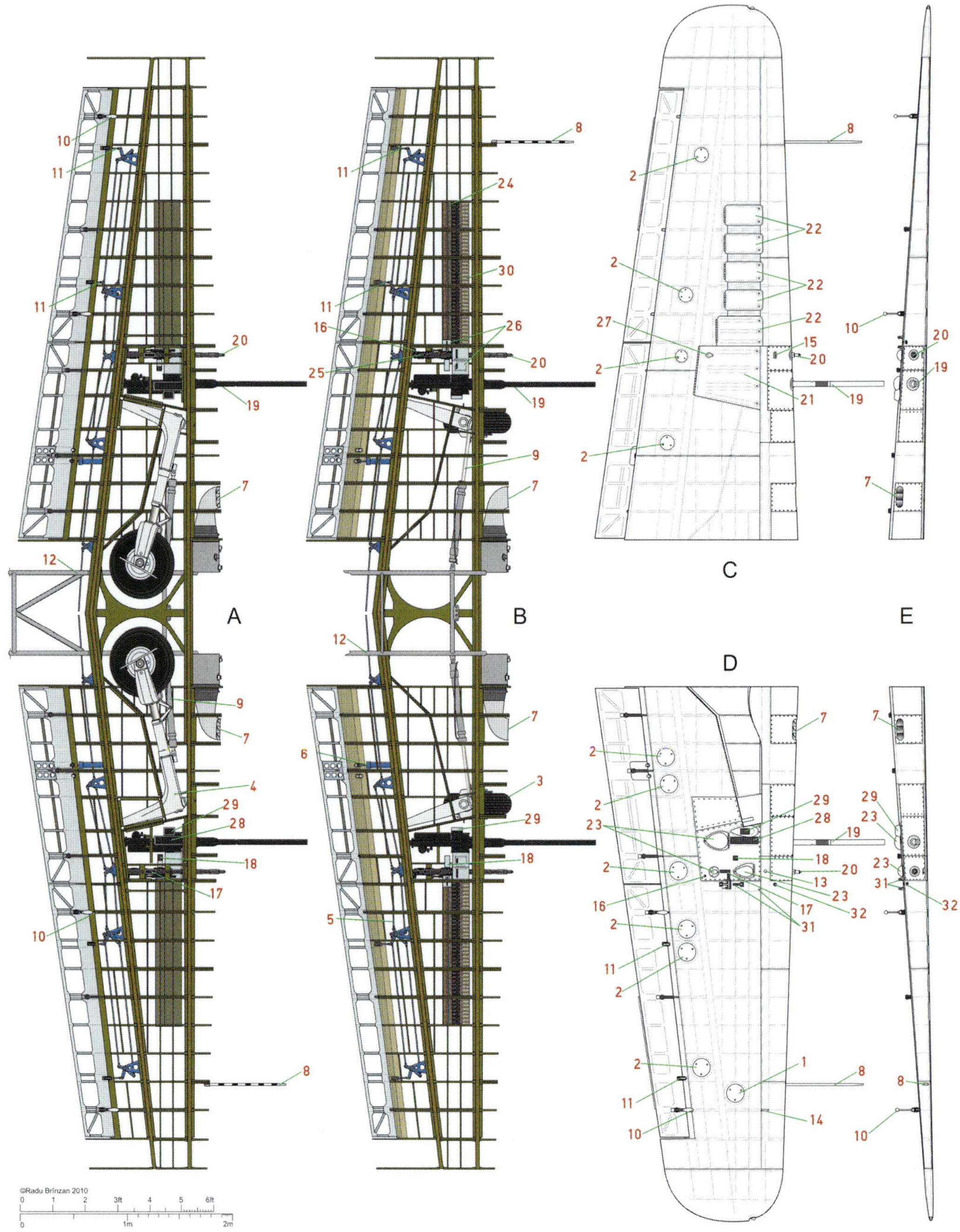

Mauser Cannon Wing: 2 x 7.92 mm FN Browning Machine Guns & 20 mm Mauser MG151/20 Cannons Used on I.A.R. 81-C Serial Numbers 301-450

A – Internal structure of the wing viewed from below;
B – Internal structure of the wing viewed from above;
C – Port wing viewed from above;
D – Port wing viewed from below;
E – Wing viewed from the front;

1. Service panel for Pitot Tube (port underside only);
2. Service panels for control linkage levers and couplings;
3. Landing gear extended;
4. Landing gear retracted – final "short leg" landing gear;
5. Control linkages;
6. Hydraulic ram for the flap;
7. Oil radiator intake;
8. Pitot Tube (port side only);
9. Hydraulic ram for the landing gear;
10. Aileron mass-balancing counterweights;
11. Aileron control linkages;
12. Internal fuselage frame (shown partially);
13. Hoist point;
14. Mooring ring;
15. Cooling vent for the gun barrel;
16. Manual gun charging toggle;
17. Spent 7.92 mm ammunition chute;
18. Spent 7.92 mm belt link chute;
19. 20 mm Mauser MG151/20 cannon;
20. 7.92 mm FN Browning machine gun;
21. Access hatch for weapons;
22. Access hatch for ammunition;
23. Domed covers for the weapon mounting and alignment fittings;
24. 7.92 mm machine gun ammunition;
25. Electrical gun-firing mechanism;
26. Ammunition feed chute;
27. Domed fairing for the 7.92 mm machine gun firing solenoid;
28. 20 mm spent ammunition chute;
29. 20 mm spent ammunition links chute;
30. 20 mm cannon ammunition;
31. Bomb and drop tank rack;
32. Drop tank fuel line intake.

Modified Mauser Cannon Wing

Modified Mauser Cannon Wing: 2 x 7.92 mm FN Browning MG & 2 x 20 mm Mauser MG151/20 Cannons Used on I.A.R.81 upgraded with the 20 mm MG151/20 cannons, 'so called' "I.A.R. 80-M"

A – Internal structure of the wing viewed from below;
B – Internal structure of the wing viewed from above;
C – Port wing viewed from above;
D – Port wing viewed from below;
E – Wing viewed from the front;

1. Service panel for Pitot Tube (port underside only);
2. Service panels for control linkage levers and couplings;
3. Landing gear extended;
4. Landing gear retracted – final "short leg" landing gear;
5. Control linkages;
6. Hydraulic ram for the flap;
7. Oil radiator intake;
8. Pitot Tube (port side only);
9. Hydraulic ram for the landing gear;
10. Aileron mass-balancing counterweights;
11. Aileron control linkages;
12. Internal fuselage frame (shown partially);
13. Hoist point;
14. Mooring ring:
15. Cooling vent for the gun barrel;
16. Manual gun charging toggle;
17. Spent 7.92 mm ammunition chute;
18. Spent 7.92 mm belt link chute;
19. 20 mm Mauser MG151/20 cannon;
20. 7.92 mm FN Browning machine gun;
21. Access hatch for weapons;
22. Access hatch for ammunition;
23. Domed covers for the weapon mounting and alignment fittings;
24. 7.92 mm machine gun ammunition;
25. Electrical gun-firing mechanism;
26. Ammunition feed chute;
27. Domed fairing for the 7.92 mm machine gun firing solenoid;
28. 20 mm spent ammunition chute;
29. 20 mm spent ammunition links chute;
30. 20 mm cannon ammunition
31. Bomb and drop tank rack;
32. Patch covering an opening for the previously-fitted 7.92 mm machine guns;
33. Drop tank fuel line intake.

©Radu Brinzan 2010

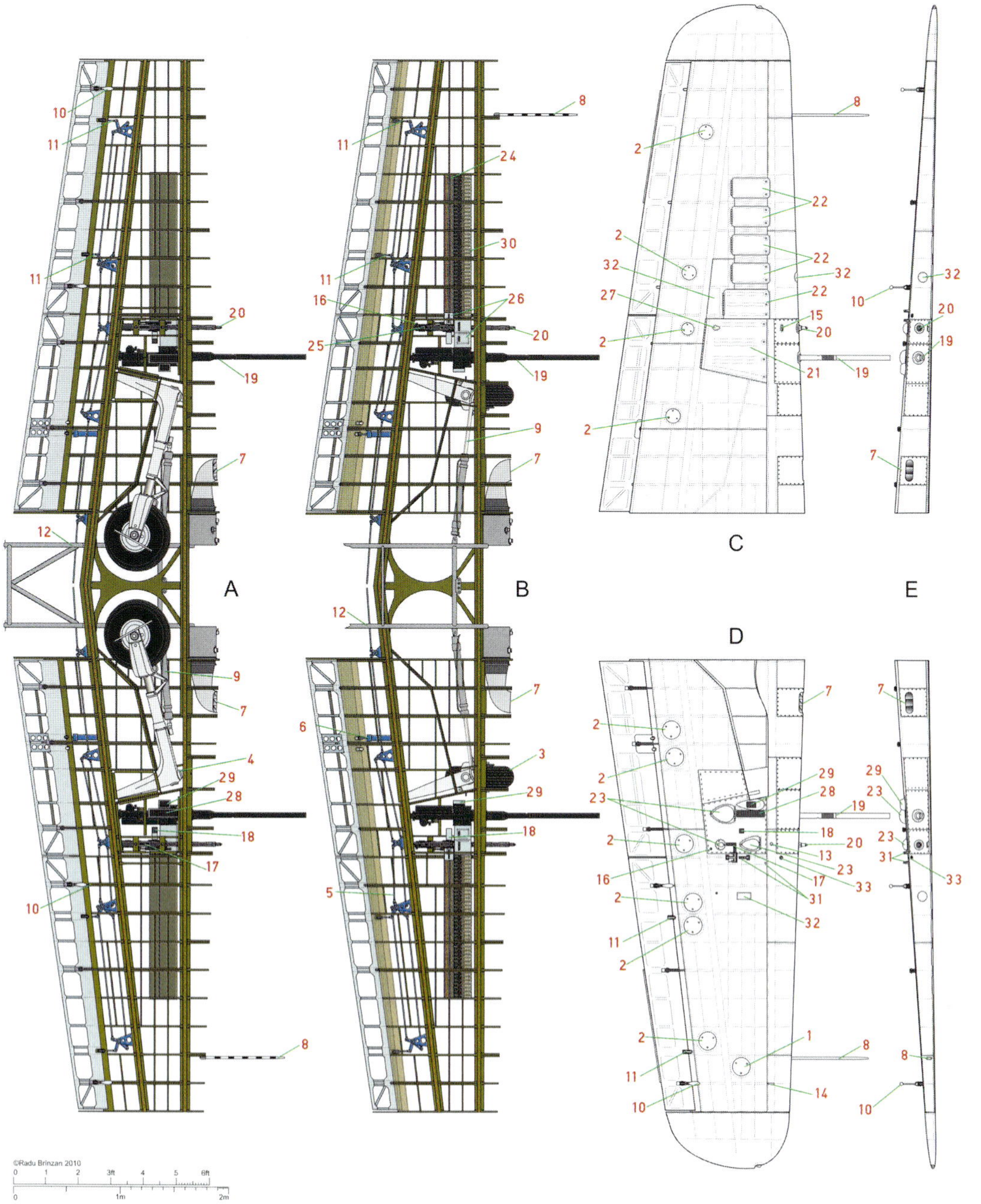

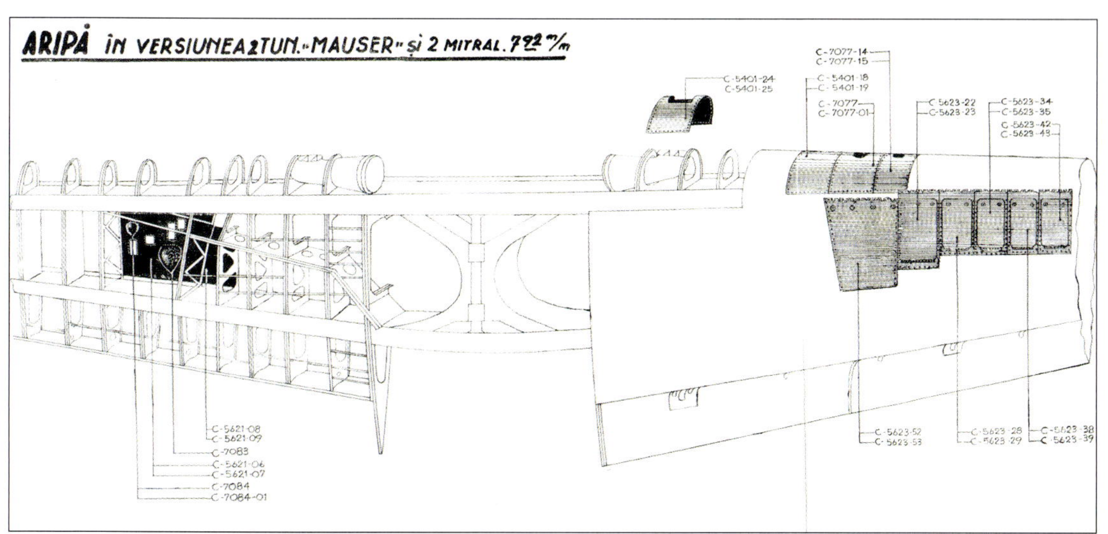

The structure and the panels of the Mauser cannon wing.
Illustration from parts list courtesy of ABC Collection

Control Surfaces

Ailerons

The ailerons of the I.A.R.80 up to No. 75 were linked to the flaps and synchronised with them to assist during take-off and landing. The normal movement range of the ailerons was from 26° up to 24° down. When the flaps were dropped, the movement range of the ailerons was from 4° up to 46° down. This photo shows the ailerons synchronised with the dropped flaps. The synchronisation mechanism between the flaps and ailerons was removed starting with I.A.R.81 No. 76 because on the I.A.R.81 the flaps were used as dive brakes. This synchronisation mechanism was never used again on any other subsequent series. Photo courtesy of ABC Collection

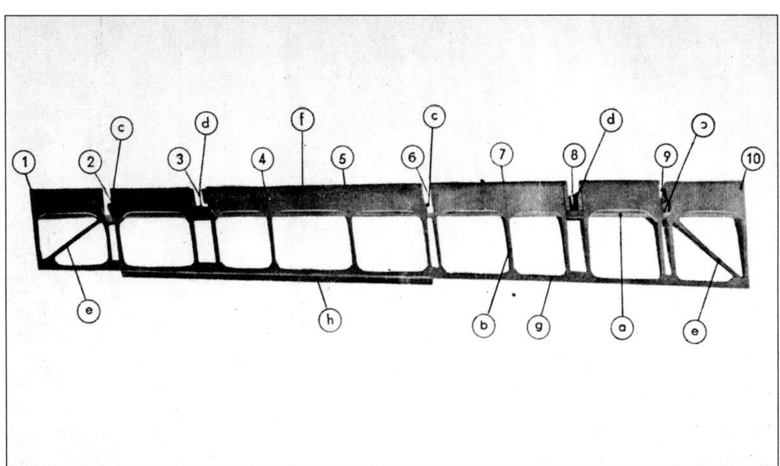

Photo of the aileron. a = Longeron tube; b = Ribs; c = Hinges; d = Control levers; e = Braces; f = Leading edge; g = Trailing edge; h = Fixed trim tab; 1–10 = Ribs. The fixed trim tabs could be adjusted on the ground across a range of 12 mm, corresponding to -40° and +40°. A Handley-Page licence was secured for the ailerons used on the I.A.R.80/81.
Photo from the technical manual courtesy of ABC Collection

Flaps

The hydraulic ram for the flap is marked 'a' in this picture. The item marked with the handwritten 'b' was the linkage that synchronised the movement of the ailerons with the movement of the flaps. This link was removed starting with I.A.R.80 No. 76. Because the flaps were hydraulically-controlled, they usually dropped when parked after a while, due to loss of hydraulic pressure.
Photo from the technical manual courtesy of ABC Collection

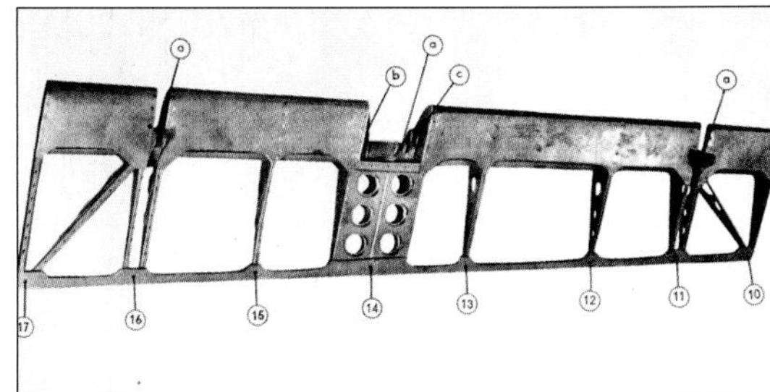

The internal structure of the flap viewed from the underside. a = Hinges; b = Flap control lever; c = Lever synchronising the flap with the aileron; 10 – 17 = Ribs. This is the early flap used up to I.A.R.80 No. 75. A new strengthened flap was introduced starting with No. 76. According to a memo regarding the strength of the flaps as dive brakes and the wing structure, issued by I.A.R. Braşov on 2 May 1941, 'In the case of aircraft intended for dive bombing, a new flap is to be built with a longeron made from chrome-molybdenum steel. This flap will be used starting with No. 151 onwards. In the case of aircraft No. 78 to 150, because of the shortage of materials, the old flaps will be used and strengthened.' This strengthened flap featured a reinforcement bar in the middle between ribs 11 and 16
Photo from the technical manual courtesy of ABC Collection

Tail Planes and Elevators

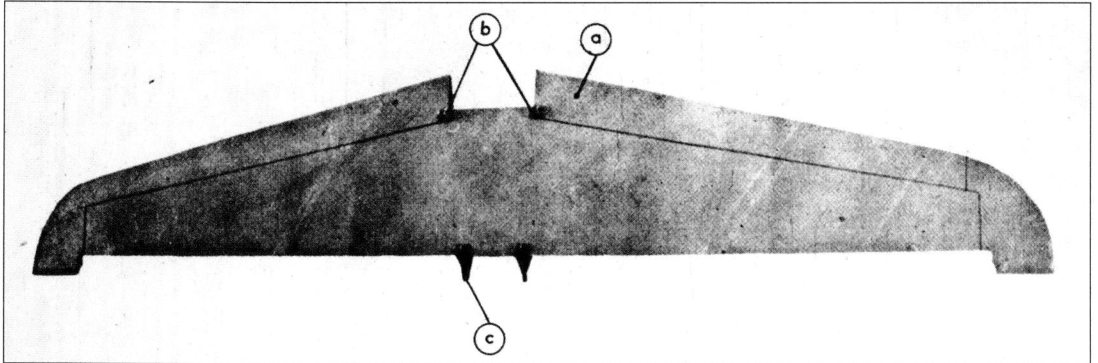

The tail plane. a = Leading edge; b = Brackets for attachment to the fuselage; c = Hinges. courtesy of ABC Collection

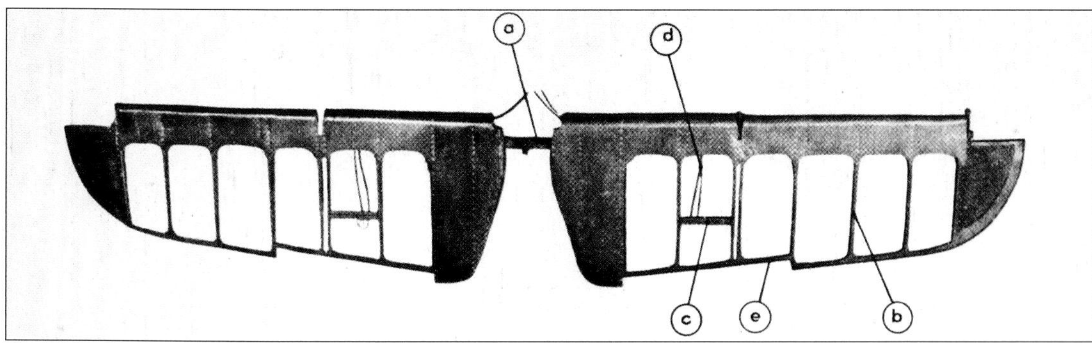

The internal structure of the elevators. a = Longeron tube; b = Rib; c = Crossbar for mounting the trim tab cable; d = Control Bowden cable for the trim tab; e = Support crossbar for the trim tab Courtesy of ABC Collection

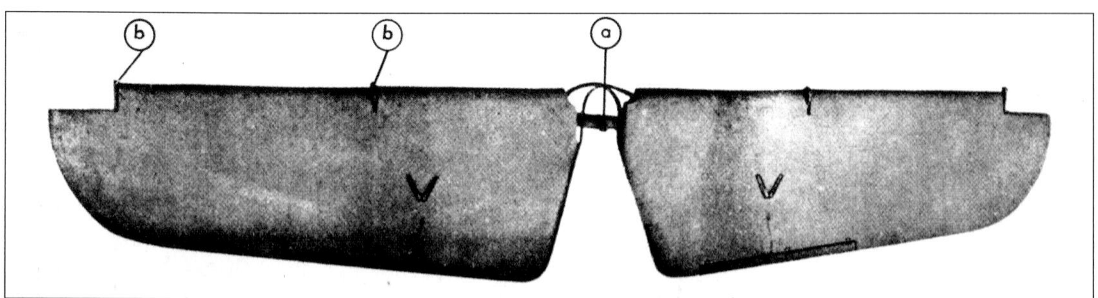

The fabric-covered elevator. The V-shaped devices visible on the fabric surfaces in this photo are zipper service panels for the trim tab control cable. a = Control lever; b = Hinges
Courtesy of ABC Collection

On 10 January 1944, I.A.R.81-C No. 398 of Esc.58Vt./Gr.7Vt. based in Pipera hit a barrack during take off while flown by Adj.Av. Ion Mălăcescu. This photo offers a very good rear view of the tail plane support strut. The support struts were initially used on the I.A.R.81 and were standardised for all types of I.A.R.80 and 81 starting with No. 241
Courtesy of ABC Collection

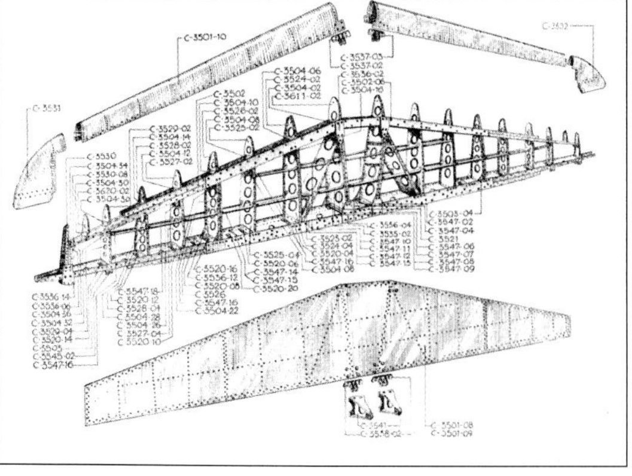

The tail plane components
Illustration from the parts list courtesy of ABC Collection

The internal components of the elevator
Illustration from the parts list courtesy of ABC Collection

Tail Fin and Rudder

The tail fin. a = Vertical (rear) spar of the tail fin;
b = Oblique spar of the tail fin; c = Leading edge;
e = Skin of the upper part of the tail fin (1 mm
Elektron); f = Dural skin (0.6 mm); g = Attachment
for the radio antenna; h = Position light location.
Photo from the technical manual courtesy of ABC Collection

The rudder components
Illustration from the parts list
courtesy of ABC Collection

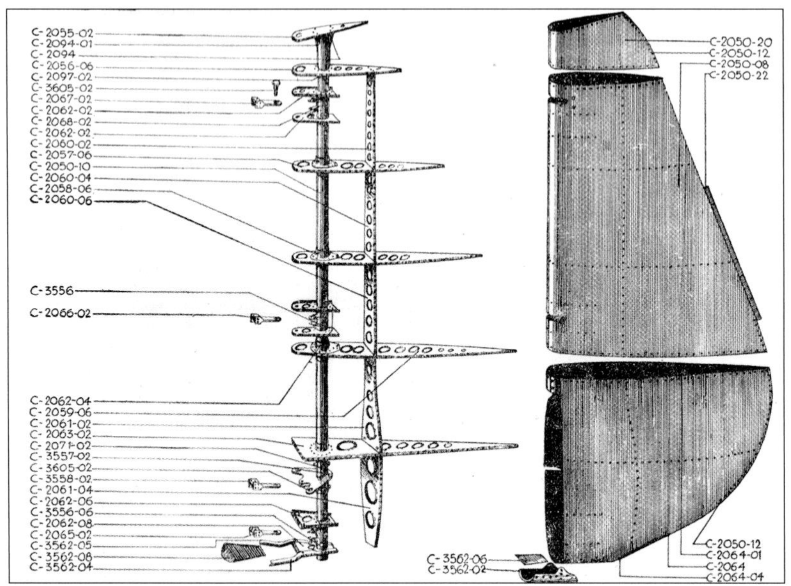

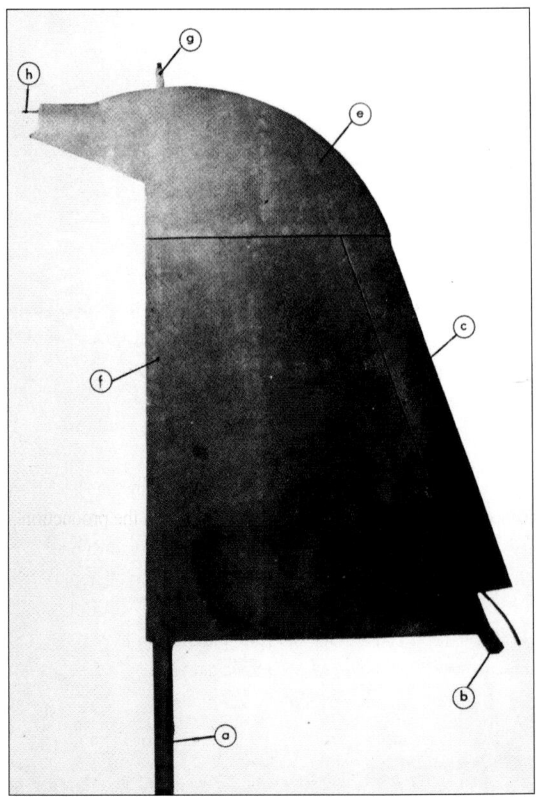

The complete tail plane and
tail fin unit viewed from the
rear. a = Fairing blending
the tail plane to the fuselage;
b = Bolt for removing the
tail fin.
Photo from the technical manual courtesy
of ABC Collection

Tail Fin and Rudder

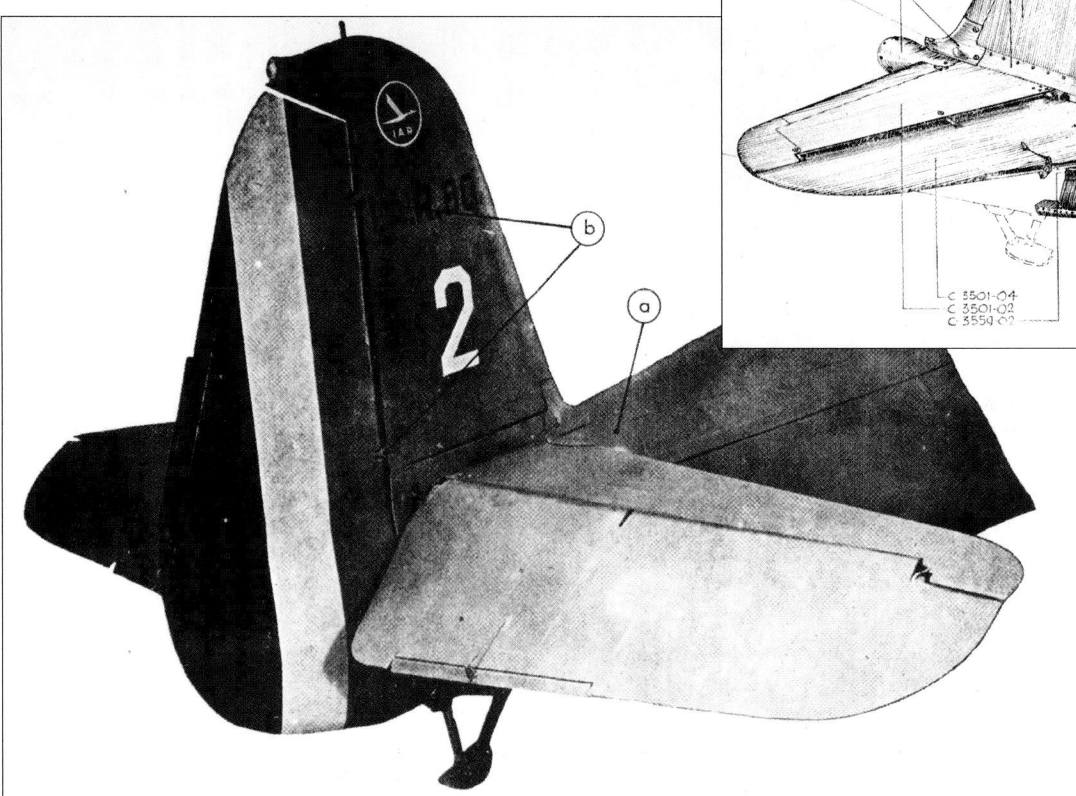

Tail unit components
Illustration from the parts list courtesy
of ABC Collection

Engine

The I.A.R.80/81 was powered by a Gnome Rhône 14K Mistral Major manufactured under licence by I.A.R. under the name of I.A.R.-14K. The licence was purchased on 21 January 1936. At the same time, a pre-production batch of 20 Gnome-Rhône 14K engines were ordered, due for delivery around May-June. They would eventually be delivered between August and November 1937. This delay postponed the manufacture of the engine under licence by I.A.R. In order to equip the aircraft on the production line, I.A.R. Braşov ordered 173 French-made Gnome-Rhône 14K II C32 on 19 November 1936, which were eventually delivered between September 1938 and April 1940. The only such engine was fitted to the I.A.R.80 prototype. Series production of the I.A.R.-14K engine began in early 1941.

Engine viewed from the right. This engine is fitted with baffles that direct the cooling air between the cooling fins of the cylinders. The horizontal pipes extending from the back of the cylinders are the exhaust pipes
Courtesy of ABC Collection

Description

Official designation	I.A.R.-14K IVc32.1000A1
Direction of rotation	Counter-clockwise, viewed from cockpit
Gear ratio	2/3
Outside diameter	1,264 mm
Length	1,844 mm
Test altitude	3,200 m
Power delivered at 3,200 m, at 2,300 RPM, at nominal boost	745 kW / 1,000 hp
Maximum cruising power	715 kW / 960 hp
Nominal power at ground level	656 kW / 880 hp
Nominal propeller RPM	1,533 RPM
Maximum RPM permitted with reduced load	2,420 RPM
Number of cylinders	14
Cylinder arrangement	Radial
Bore	146 mm
Stroke	165 mm
Displacement	38,67 litres
Compression ratio	5:5.1
Carburettor	Zenith - Stromberg NAR 130 RGSL
Fuel consumption	295 +15 / -7 g/hp per hour
Oil consumption	10±5g/hp per hour
Magnetos	2 Bosch GE 14L11
Spark plugs	28 x Bosch DM175ET6
Engine start	Electrical: Bendix Elipse 11A / manual crank
Fuel pumps	2 x I.A.R. Model IV
Valve train	Overhead valves
Supercharger	Single-speed, centrifugal

The engine cooling flaps were hydraulically operated. This photo of the engine rear shows the supercharger. The black cylinder at the top is the electrical engine starter
Courtesy of ABC Collection

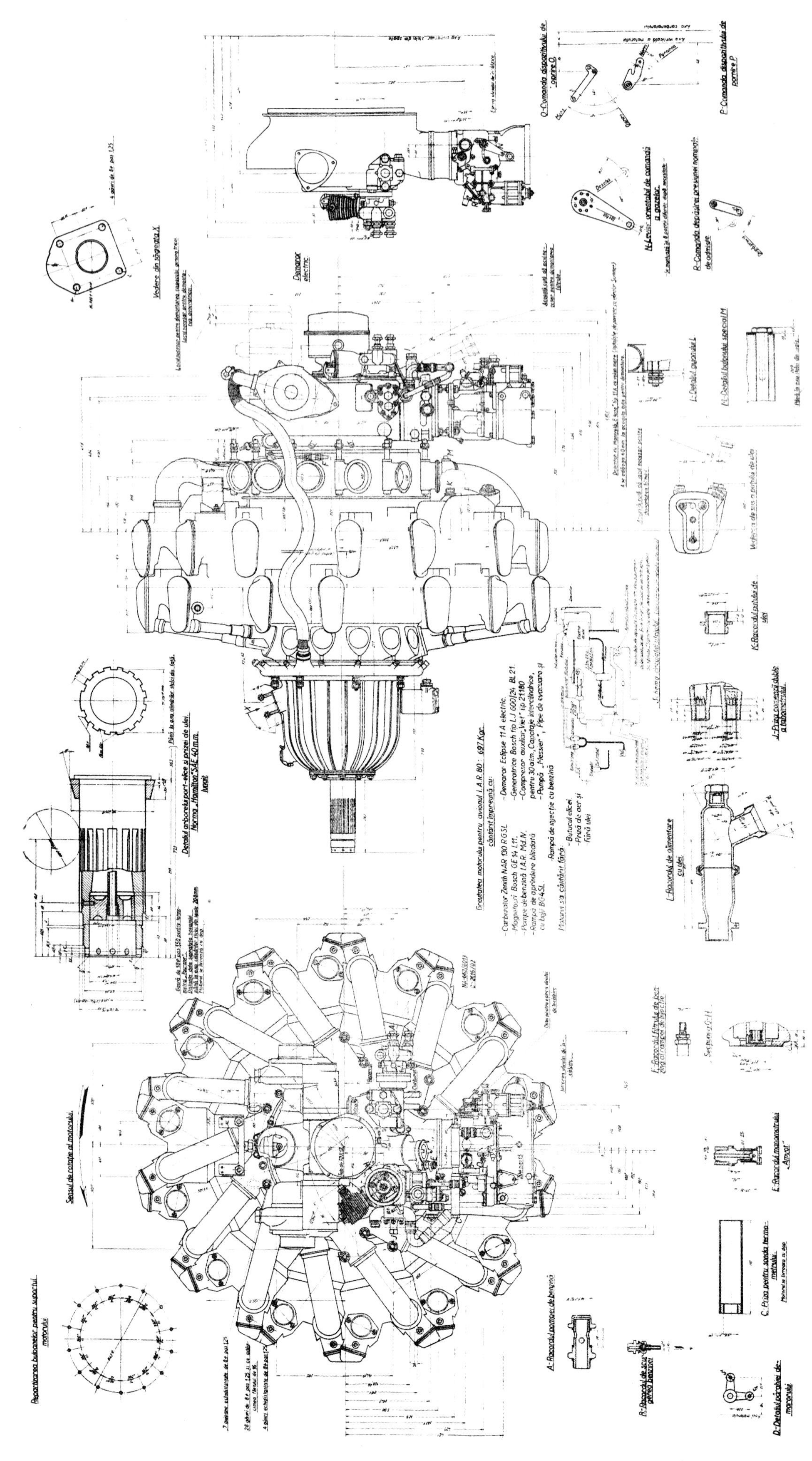

Cross section engine drawing
Courtesy of ABC Collection

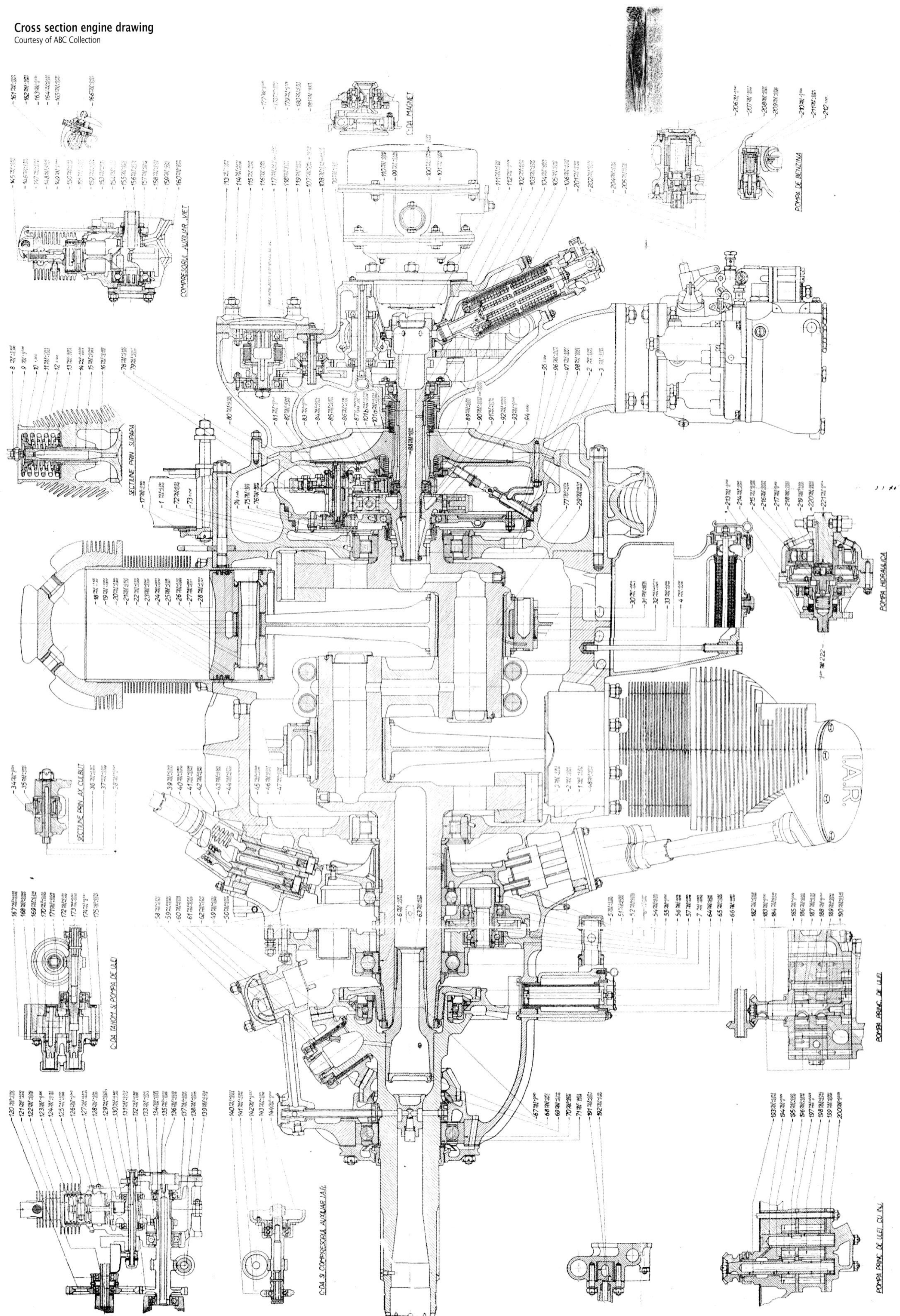

Head-on photo of the I.A.R.-14K IVc32 1000A engine
Photo courtesy of ABC Collection

Propeller hub. 6 = Gear train for propeller control; 14 = Access panel for lubrication; 15 = Axle of the endless screw; 18 = Bushing for blade bearer; 19 = Intermediate bearer; 26 = Backplate; 17 = Hub lubricator for the main gear train.

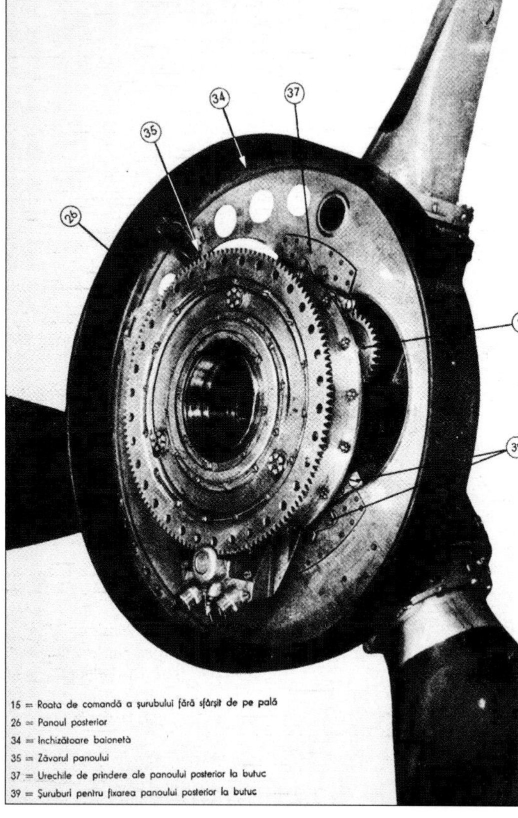

Backplate and propeller pitch control mechanism. 15 = Control wheel of the endless screw on the blade; 26 = Backplate; 34 = Bayonet lock; 35 = Backplate lock; 36 = Brackets for securing the backplate to the axle; 39 = Screws for securing the backplate to the axle.
Courtesy of ABC collection

Frontal view of the propeller
Note the VDM manufacturer logo. Courtesy of ABC collection

Rear view of the propeller showing the pitch control mechanism.
Courtesy of ABC collection

Ignition wiring
Courtesy of ABC Collection

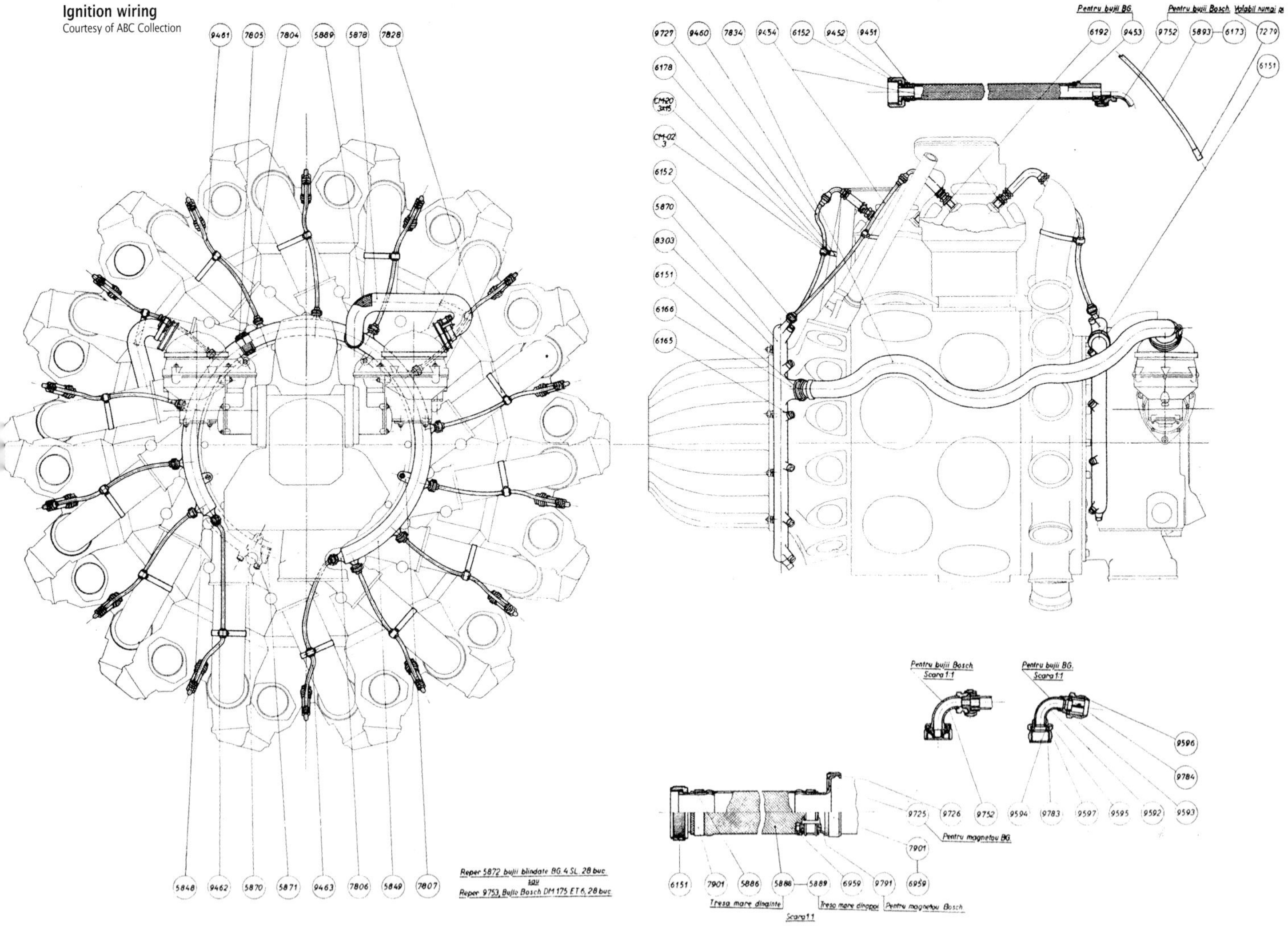

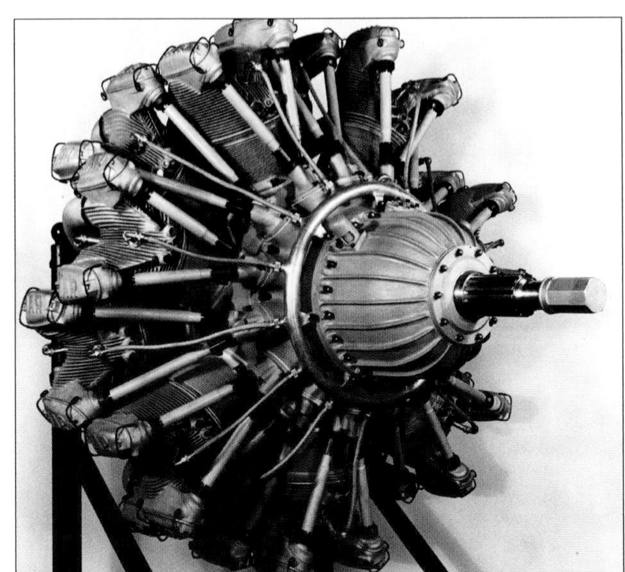

Three-quarter frontal view of the engine.
Courtesy of ABC Collection

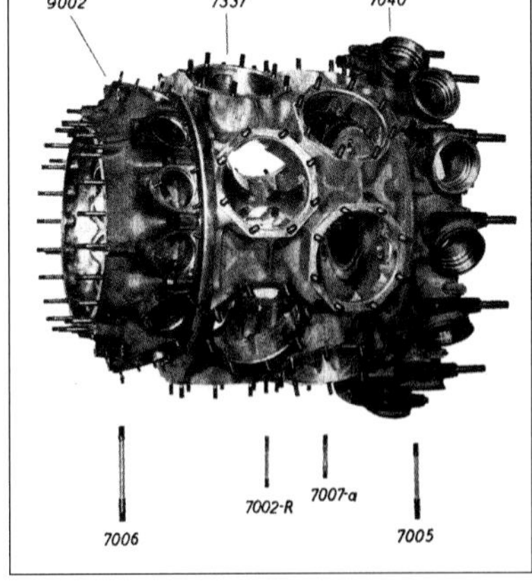

Engine block with the intake manifold bolted at the rear. The bolts protruding on the right side in this picture are the mounting bolts that secured the engine to the bearer dampers. Courtesy of ABC Collection

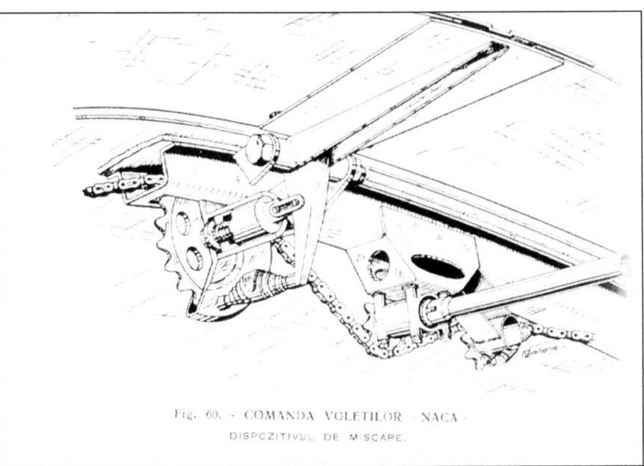

This perspective drawing shows the cogs and the bicycle chain that synchronised them. Intermediate curved rectangular panels fitted between the movable flaps inside deep channels specially provided on either side of each flap. As the flaps moved, these panels moved in tandem with them and thus formed a continuous variable 'skirt' around the rear edge of the cowl
Courtesy of ABC Collection

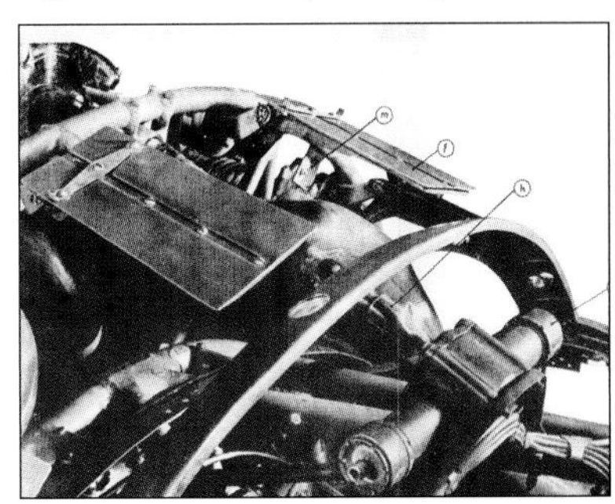

Engine cooling flaps. g = Rack and pinion ram; h = Torsion axle for moving the chain; f = NACA cooling flap; m = Bracket securing the exhaust collector to the rear ring of the NACA engine cowl ring.
Courtesy of ABC Collection

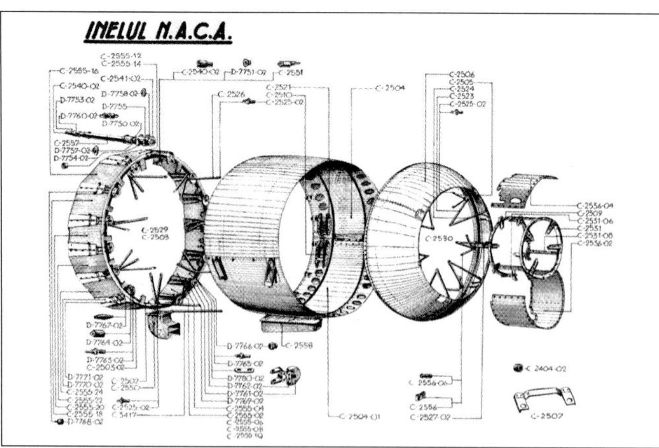

Engine cowl components
Courtesy of ABC Collection

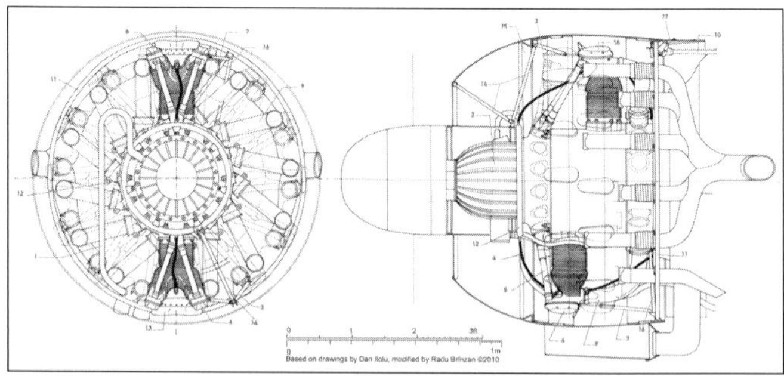

Cowl assembly: 1 = Cooling pipe for the propeller pitch gear; 2 = Reduction gear box; 3 = Circumferential supports for the front cowl ring. These linked the front of the cylinder heads and the front cowl mounting ring. There were seven of these, one for each cylinder; 4 = Sparkplug cables; 5 = Pushrods; 6 = Front-row cylinder; 7 = Circumferential supports for the rear cowl frame. These linked the rear of the cylinder heads and the rear cowl frame. There were seven of these, one for each cylinder; 8 = Back-row cylinder; 9 = Fuel mixture intake; 10 = Cowl flaps control; 11 = Rear cowl frame. The cooling flaps were mounted onto this frame. This frame also provided support for the removable cowl panels; 12 = Sparkplug wiring loom; 13 = Rubber air-deflection baffle; 14 = Radial supports for the front cowl ring. These linked the engine block and the mounting ring of the front cowl. There were seven of these, one for each cylinder; 15 = Mounting ring of the front cowl; 16 = Radial supports for the rear cowl cowl. These linked the engine block to the rear cowl frame. There were seven of these, one for each cylinder; 17 = Cooling flaps.

An extended duct was added to protect the engine from dust, but it was discovered to be ineffective and the engine continued to be very susceptible to dust damage. An air filter for the supercharger intake controlled from the cockpit was introduced beginning with No 241. Starting with No. 310, the shutter cable was linked to the landing gear and the shutter was automatically closed when the landing gear was extended. This is an enlarged photo of the filter. The shutter at the front of the filer is closed. The operating cable of this shutter leads from a tubular mount fitted between the back of the engine cowl and the front of the cooling flap to a lever above the shutter door. The filter is held in place by hooked brackets extending below the cowl.
Courtesy of ABC Collection

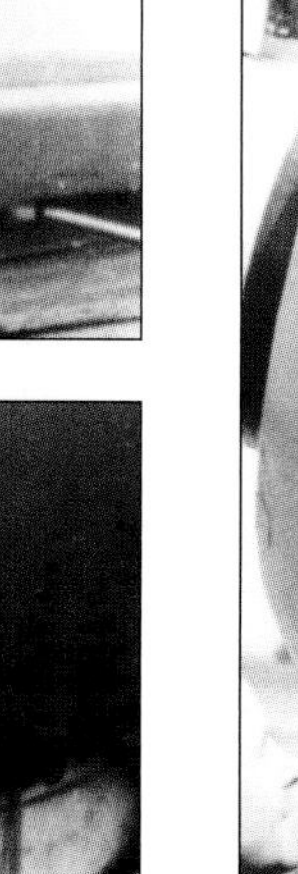

I.A.R.81-C No. 336. The diagonal support braces connecting the rear of the domed forward cowl to the front of the cylinders are visible
Courtesy of ABC Collection

The I.A.R.14K was built in 3 versions, each bringing an improvement in performance:

- I.A.R.-14K IIc32, 671 kW / 900 hp take-off output and 648 kW / 870 hp nominal output. This engine was fitted only to the prototype. Manufacturing series began with No. 1001;
- I.A.R.-14K IIIc36, 745 kW / 1,000 hp take-off output and 693 Kw / 930 hp nominal output. This engine was fitted to I.A.R.80 No 1. Manufacturing series began with No. 4001;
- I.A.R.-14K IVc32 1000A, 745 kW / 1,000 hp take-off output and 715 kW / 960 hp nominal output. Manufacturing series began with No. 5001. I.A.R.-14K IVc32 1000A1 was a lighter version of the I.A.R.14K engine and was the type used on the majority of aircraft.

The serial numbers of each engine fitted to each aircraft were meticulously recorded by the factory and the workshops. Although it is possible to compile a list matching the engine serial numbers with the aircraft they were fitted to, such a list would exceed the purpose and available space of this book since all aircraft were fitted with a number of engines throughout their service. However, a study of these serial numbers points out that many engines were used a number of times on different airframes following overhaul and refit.

The schedule for engine maintenance was as follows:
- After 10 hours of flight: check the engine mount; check the engine cowl, supports, fasteners and cooling flaps;
- After 25 hours of flight: check engine controls, check engine mount; check propeller fittings and alignment
- After 100 hours of flight: partial overhaul of the engine
- After 300 hours of flight: general overhaul of the engine.

In reality, due to the excessive strain that the engine was subjected to, the time between overhaul of the engine was 30 hours on average. The I.A.R.14K engine was a weak point of the I.A.R.80 because of the high oil consumption and short time between overhaul.

Landing Gear

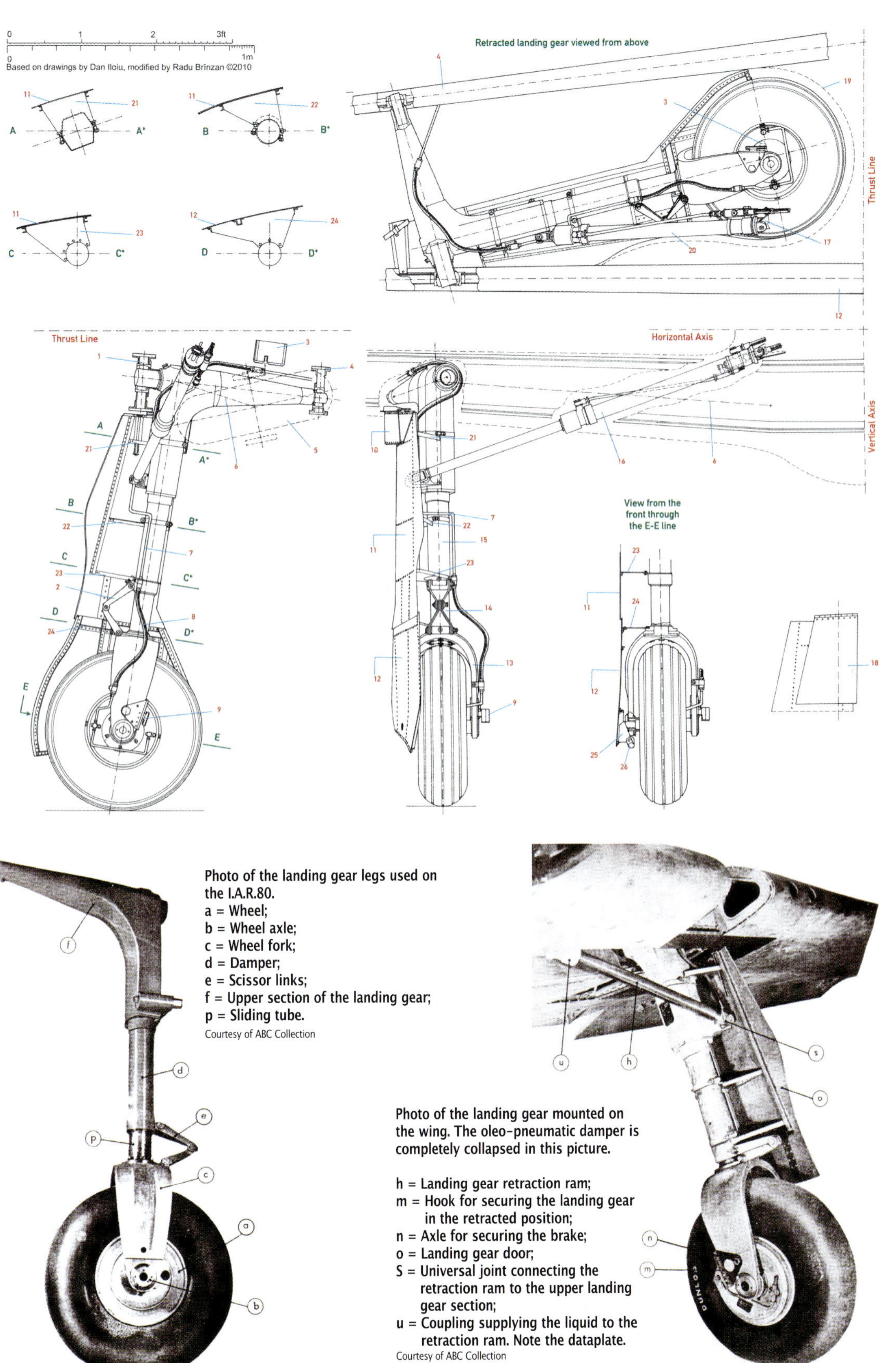

This is a drawing of the early landing gear used on all airframes up to No. 75. From No. 76 onwards, the oleo was shortened by 60 mm.

1 Forward spar
2 Scissor links
3 Landing gear up-lock
4 Rear spar
5 Position of the wheel when retracted
6 Position of the landing gear leg axis when retracted
7 Rigid air line for the brakes
8 Flexible air line for the brakes
9 Landing gear lock hook
10 Hinged top cover for the landing gear
11 Upper landing gear door
12 Lower landing gear door
13 Landing gear fork
14 Oleo-pneumatic damper
15 Landing gear leg
16 Hydraulic retraction strut
17 Retraction strut attachment pivot
18 Upper landing gear door (Only on the first 20 airframes for a short time. Replaced later with the 'hinged top cover', item 10 above, on all airframes)
19 Wheel well outline
20 Hydraulic retraction strut in the retracted position
21 Top mounting bracket of the upper landing gear door
22 Middle mounting bracket of the upper landing gear door
23 Bottom mounting bracket of the upper landing gear door
24 Top mounting bracket of the lower landing gear door
25 Bottom mounting bracket of the lower landing gear door
26 Mounting bar for the lower landing gear door - this was welded to the landing gear fork

Photo of the landing gear legs used on the I.A.R.80.
a = Wheel;
b = Wheel axle;
c = Wheel fork;
d = Damper;
e = Scissor links;
f = Upper section of the landing gear;
p = Sliding tube.
Courtesy of ABC Collection

Photo of the landing gear mounted on the wing. The oleo-pneumatic damper is completely collapsed in this picture.

h = Landing gear retraction ram;
m = Hook for securing the landing gear in the retracted position;
n = Axle for securing the brake;
o = Landing gear door;
S = Universal joint connecting the retraction ram to the upper landing gear section;
u = Coupling supplying the liquid to the retraction ram. Note the dataplate.
Courtesy of ABC Collection

Wheels

The I.A.R.80/81 used 635 x 190 mm tyres made by Dunlop, BF Goodrich or Pirelli. The tyre load was 1,250 kg for Dunlop tyres and 1,500 kg for BF Goodrich tyres with a tyre pressure of 3.15 kg/cm2.

Dunlop tyres were manufactured under licence by Fabrica de Anvelope Floreşti and BF Goodrich tyres were made under licence by Fabrica de Anvelope Banloc. Late in the war, the quality of the tyres manufactured in Romania by Fabrica de Anvelope Banloc under a BF Goodrich licence became increasingly poor. They were the subject of many complaints from the pilots and maintenance crews as well as the cause of a number of accidents. In December 1943, I.A.R. Braşov took a lawsuit against Fabrica de Anvelope Banloc, but the quality of the tyres continued to be a source of difficulties.

The final type of tyre used on the I.A.R.80/81, also made by BF Goodrich, featured a grooved pattern. Courtesy of ABC Collection

The brakes used on the I.A.R.80/81 were a Messier design and were pneumatically operated.
a = Brake plate;
b = Brake pads;
c = Control pistons for the pads,
d = Wheel drum;
e = Axles for adjusting the callipers (contact across the entire surface);
f = Axle for adjusting the callipers (proximity or distance of the callipers from the drum).
Courtesy of ABC Collection

Landing Gear Doors

The landing gear leg was fitted with an etched metal plate containing maintenance instructions. The instructions are, from the top: 'MAINTENANCE', 'KEEP THE ROD OF THE PISTON CLEAN AND WELL-OILED', 'NORMAL INFLATION HEIGHT__m/m to__m/m measured from the bottom of the cylinder to the shoulder of the piston', 'In case of abnormal inflation, consult the maintenance leaflet', 'NEVER INFLATE WITH OXYGEN', 'Licensed by Messier, BUILT BY SEMAT S.A. Bucharest'
Drawing by Radu Brînzan

Enlarged side view of the crashed I.A.R.80 No. 2. This illustrates the doors used with the early 'long' landing gear. Note the pressure instruction stencil on the bottom section. Piteşti Military Archives
Courtesy of ABC Collection

Lateral view of the final 'short' landing gear door. In this photo, the landing gear is compressed under the weight of the fully-loaded aircraft which creates a 'step' in the outline of the doors at the front and rear. When the landing gear extended fully, such as when it is retracted, this 'step' was no longer present. Courtesy of ABC Collection

Internal view of the final 'short' landing gear door showing the mounting brackets that secured the landing gear doors to the landing gear legs. Courtesy of ABC Collection

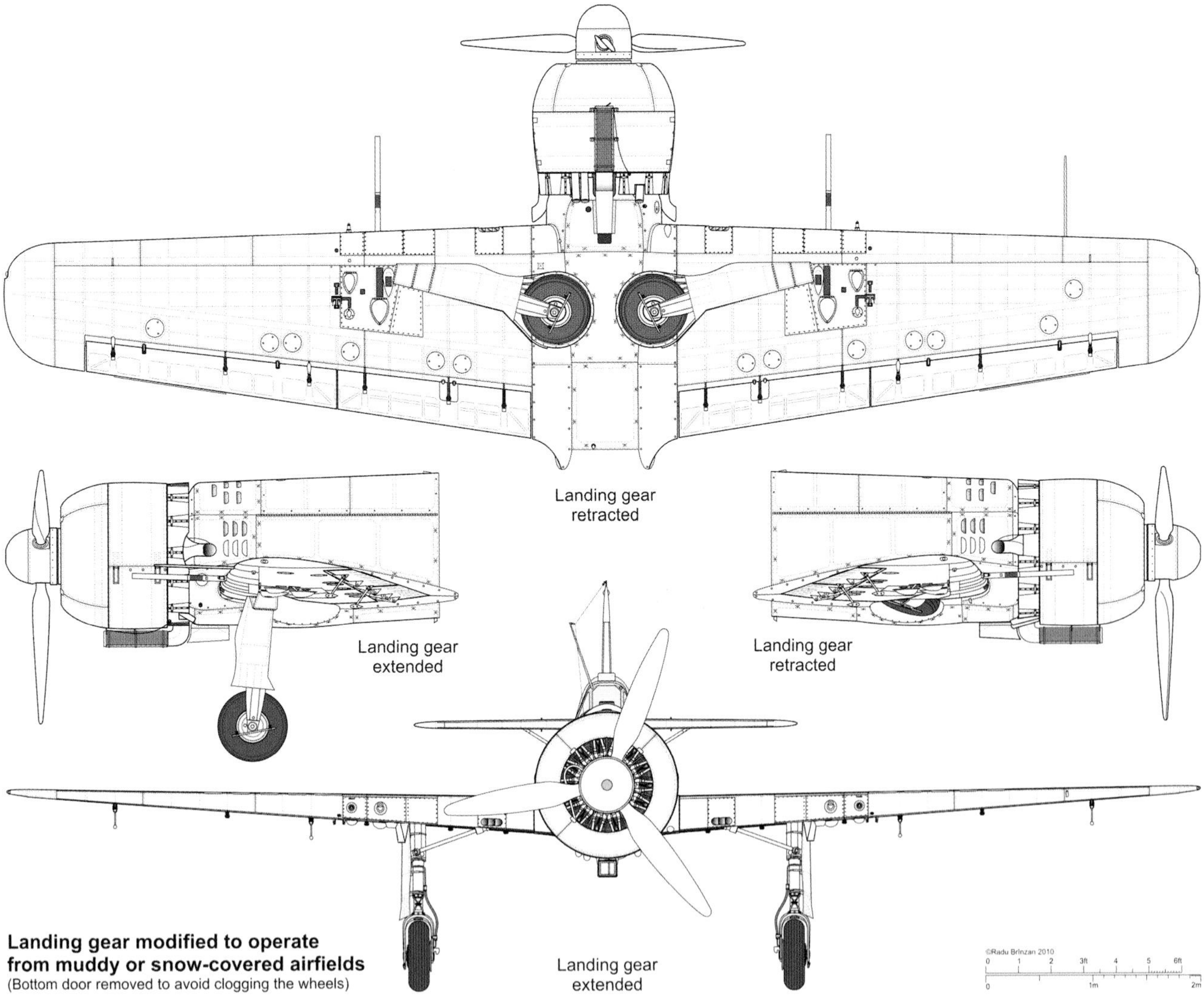

Landing gear retracted
Landing gear extended
Landing gear retracted
Landing gear modified to operate from muddy or snow-covered airfields
(Bottom door removed to avoid clogging the wheels)
Landing gear extended
©Radu Brînzan 2010
0 1 2 3ft 4 5 6ft
0 1m 2m

Tail Skid

a
= Carenaj fixat cu şuruburi „Elastic stop"

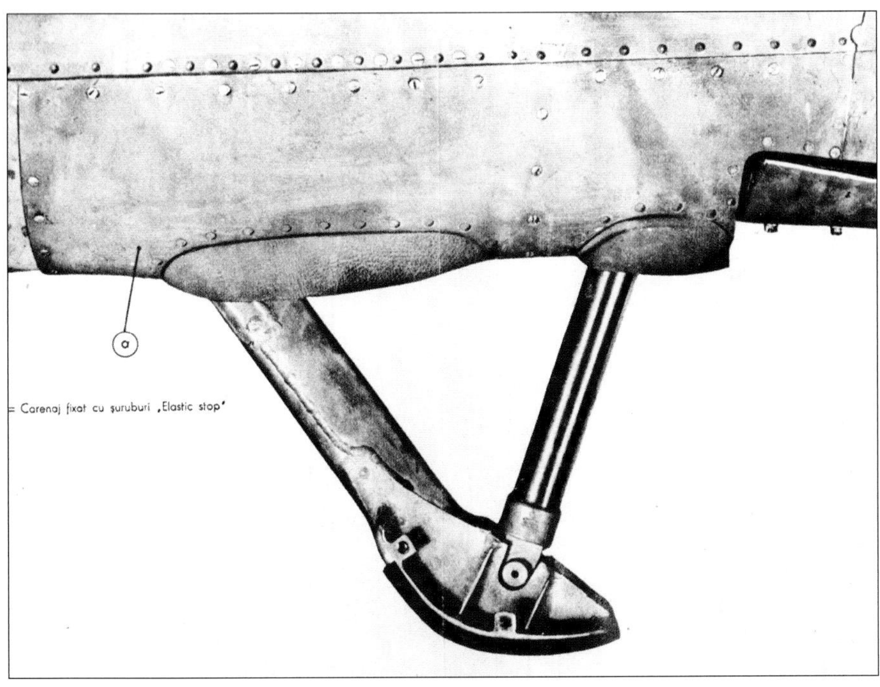

The I.A.R.-UT14 tailskid was used on all types of I.A.R.80/81. This arrangement meant that the aircraft could only operate from grass strips.
a = Fairing secured with self-locking screws.
Courtesy of ABC Collection

This illustration shows the attachment points of the tail skid and the components.
a = Tail skid tube;
b = Tail skid damper;
c = Streamlined fairing of tube a;
d = Axle mounting the tube to the fuselage;
e = Mounting bracket for the damper;
f = Universal joint;
g = Skid plate. Courtesy of ABC Collection

d
c
a
b
e
g
f
a = Tubul bechiel
b = Amortizorul bechiel
c = Carenajul tubului a
d = Axul de prindere a tubului a la fuselaj
e = Ureche pentru prinderea amortizorului
f = Cordan
g = Patina bechiei

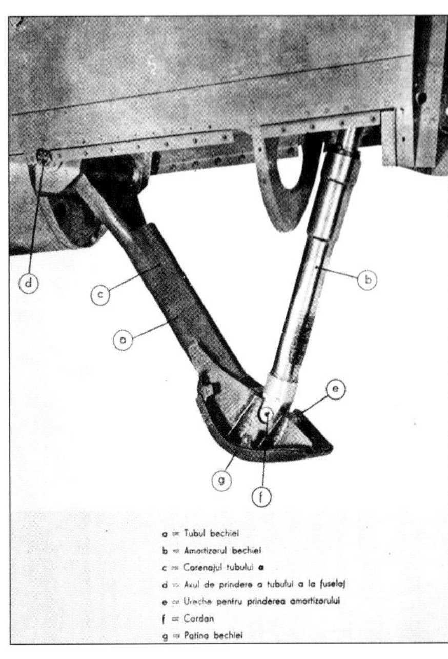

Systems

Fuel System

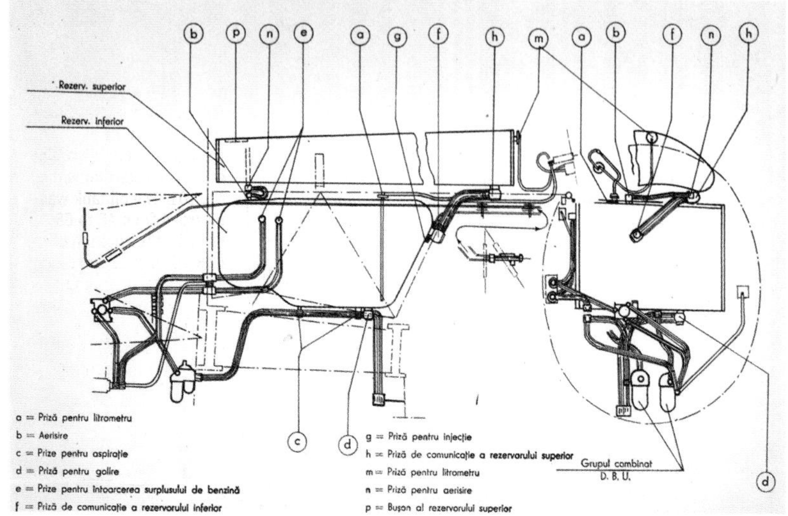

Diagram of the two fuel tanks and their connections. Grupul combinat D.B.U. = Combined D.B.U. Group; Rezerv. superior = Upper tank; Rezerv. inferior = Lower tank; a = Connection to the fuel gauge; b = Breather; c = Supply connections; d = Fuel drain connection; e = Connection for the return of fuel surplus; f = Communicating connection of the lower tank; g = Choke connection; h = Communicating connection of the upper tank; m = Connection to the fuel gauge; n = Breather connection; p = Upper fuel tank cap. Courtesy of ABC Collection

Drop Tanks

On 26 August 1941, a decision was taken to test the use of drop tanks on the I.A.R.80 in order to increase the range of the fighter. Modifications had to be made to the fuel system with the addition of special plumbing in the wings connecting with the fuel system. The first type capable of carrying drop tanks was the I.A.R.80-B, starting with No. 201. At first, the decision was taken to use 100-litre drop tanks made from compressed cardboard captured from the Russians. The Russian 100-litre drop tanks were called ПЛГБ=подвесной легкий гофрированный бак - (PLGB = Suspended light corrugated tanks). On 23 November 1943, SSA placed an order with Fabrica de Celuloză Zarnești for 1,600 drop tanks made of compressed cardboard. These drop tanks performed well in tests. On 11 March 1944, the factory refused to manufacture any further drop tanks and IAR Brașov began to manufacture drop tanks made of Elektron sheet. Afterwards, SSA received a large cache of captured Russian drop tanks that were subsequently sent to IAR Brașov where 450 were selected and adapted for use with the I.A.R.80.

The drop tanks could be jettisoned via the electrical panel on the main instrument panel with a switch marked 'Bombe – Mitraliere' (Bombs - Machine Guns) that was also used to release the bombs on the I.A.R.81. This switch changed the function of the machine gun button on the control switch thus allowing it to discharge either the bombs or the machine guns. Aircraft were not allowed to perform any aerobatics while carrying drop tanks and pilots were advised to jettison the drop tanks before entering combat. The drop tanks could not be used immediately after take-off because the auxiliary fuel pump that operated them supplied a larger amount than that required by the engine and the surplus fuel was transferred to the main fuel tank. Therefore, the fuel tanks could not be operated until around 40 minutes into the flight when the fuel from the upper main tank was expended and enough room was available in the main fuel tank to allow for the surplus to be circulated from the drop tanks. The drop tanks had a capacity of 100 litres but at first they were only filled with 50 litres of aviation fuel because the early wing-mounted bomb racks that they were attached to could only carry a maximum payload of 50 kg. Starting with I.A.R.80-C No. 241, the wing carriers were strengthened to carry payloads of 100kg.

Lateral view of a Polikarpov I-153 Chaika carrying 100-litre PLGB drop tanks. These are the only known type of Russian 100-litre compressed cardboard drop tanks. Courtesy of Viktor Kulikov

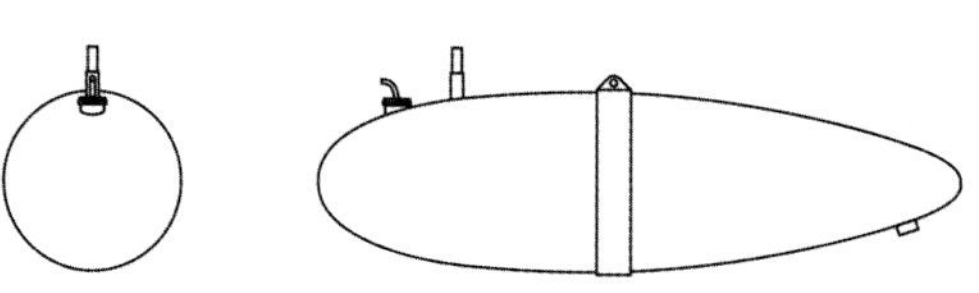

100-litre compressed cardboard Russian PLGB Drop Tank

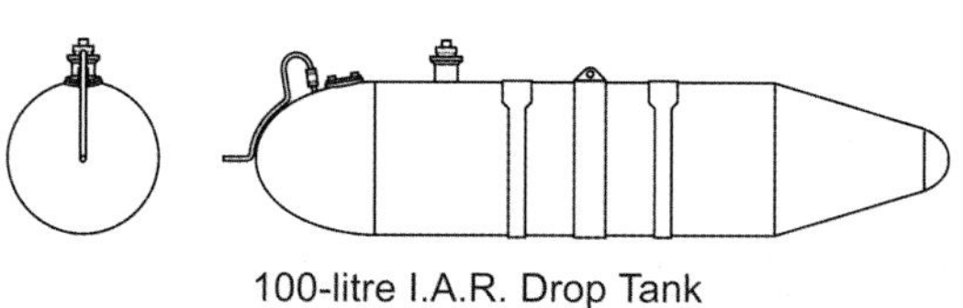

100-litre I.A.R. Drop Tank

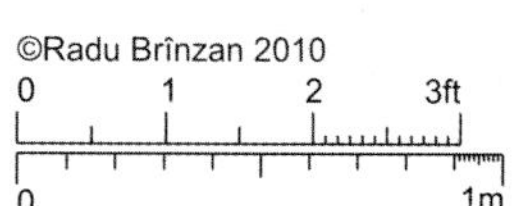

©Radu Brînzan 2010

Oil System

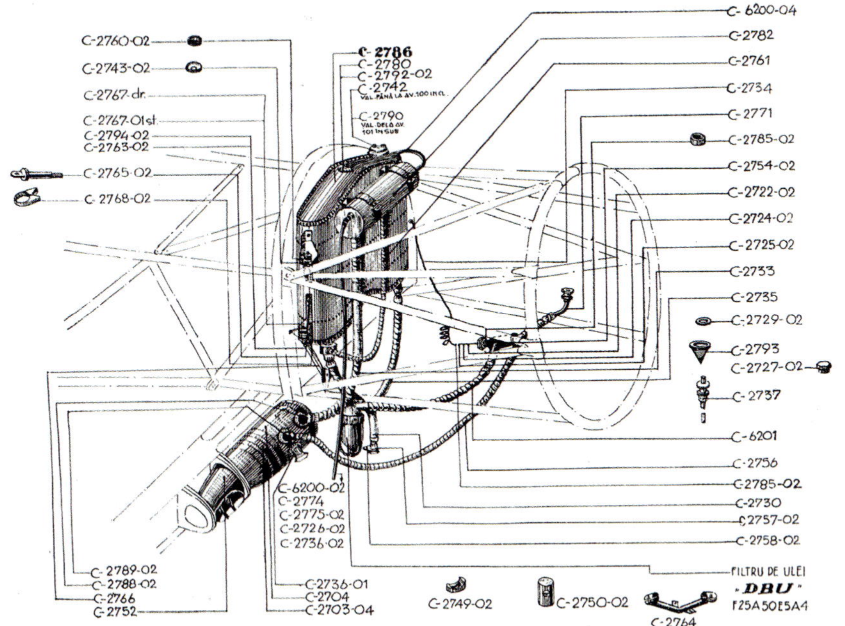

A perspective view of the increased capacity oil system with one oil cooler introduced starting with No. 101. The oil tank was enlarged from 48 to 68 litres by the addition of a cylindrical supplementary tank. A second oil cooler was introduced starting with No. 251 to aid oil cooling and was used on all subsequent airframes
Courtesy of ABC Collection

Hydraulic System

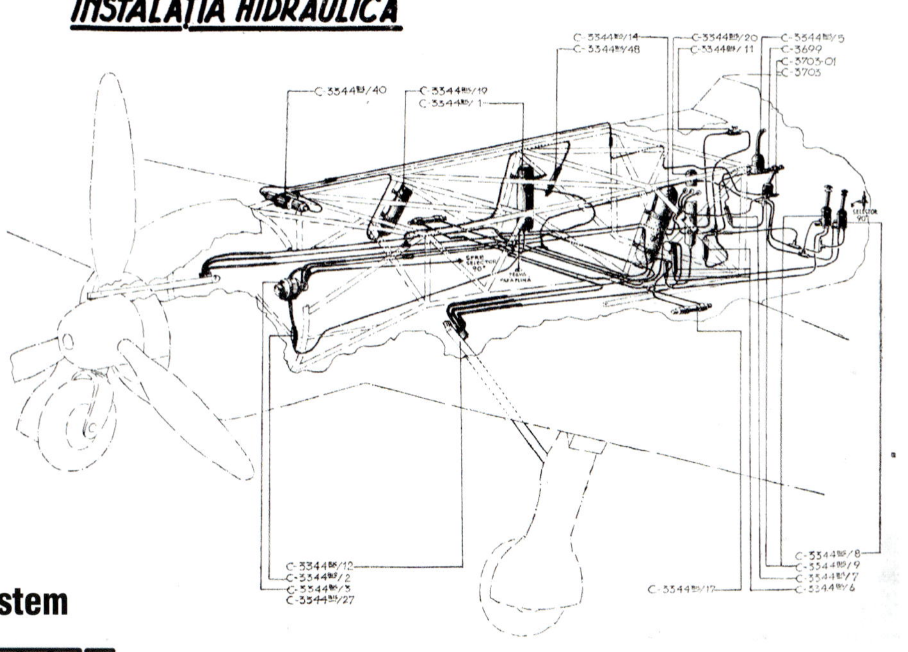

Perspective view of the hydraulic system.
Courtesy of ABC Collection

Electrical System

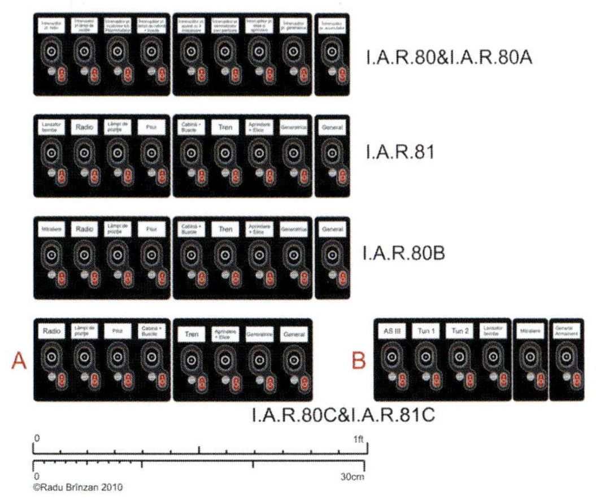

The electrical system was protected by circuit breakers. The purpose of each circuit breaker was identified by a label affixed above the corresponding button. Along with the main circuit breakers [A] mounted on the right-side diagonal shelf, the cannon-armed aircraft were equipped with a supplementary set of circuit breakers [B]. These were mounted on an angled bracket on the right sidewall of the cockpit in the place previously occupied by the breathing mask storage.

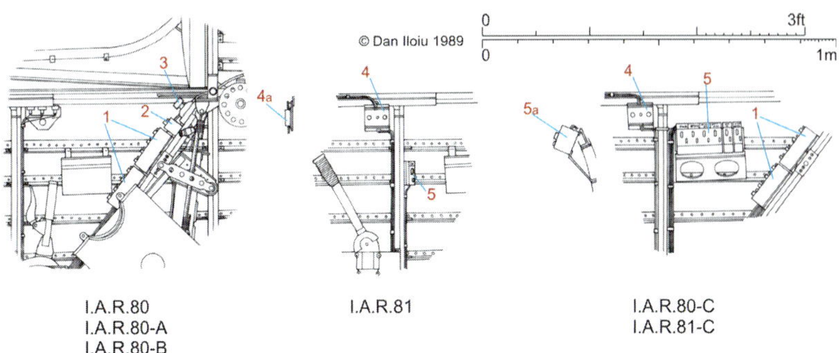

The electrical system was protected by a set of circuit breakers.
1 = Circuit breakers;
2 = Master circuit breaker;
3 = Master electrical switch;
4 = Wing bomb carrier distributor box;
4a = Side view of wing bomb carrier distributor box,
5 = Additional circuit breakers;
5a = Side view of the additional circuit breakers and their support bracket

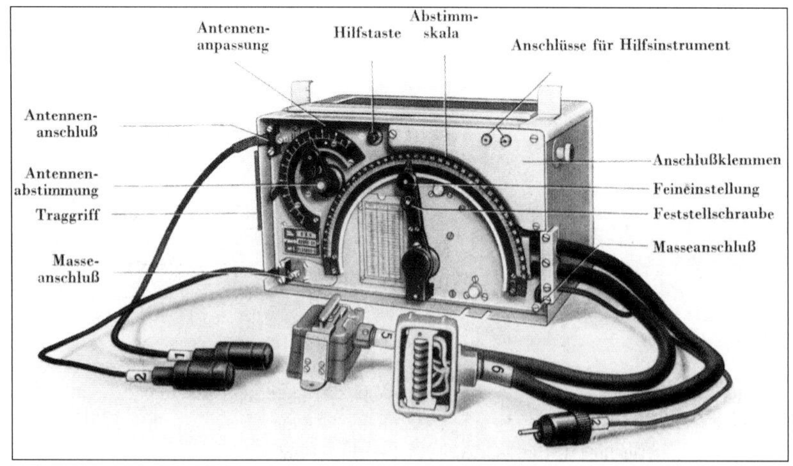

Photo of the S. 6b Transmitter. Courtesy of Günter Hütter

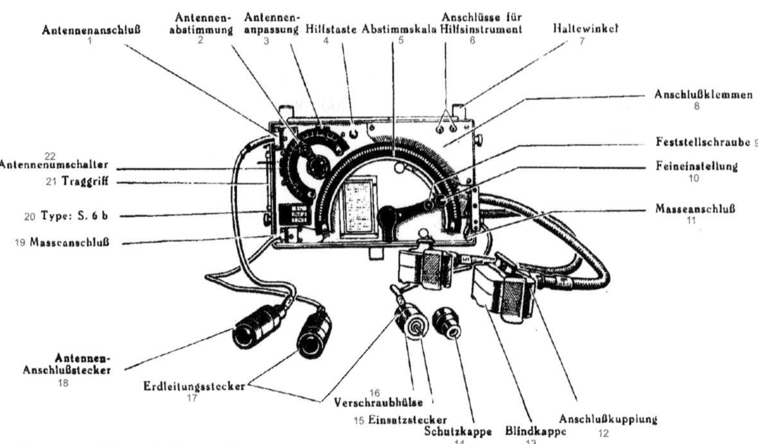

Diagram of the S. 6b Transmitter.

1 = Antenna connection;
2 = Antenna tuner;
3 = Antenna adjustment;
4 = Assistance button;
5 = Tuning scale;
6 = Sockets for auxiliary instrument;
7 = Hanging bracket;
8 = Connection pins;
9 = Frequency-setting knob;
10 = Fine-tuning knob;
11, 19 = Case connection;
12 = Connection plug;
13 = Cover cap;
14 = Protective cap;
15 = Input plug;
16 = Threaded connection case;
17 = Ground conductor plug;
18 = Antenna connection plug;
20 = S. 6b Transmitter;
21 = Carrying handle;
22 = Antenna commuting switch

Courtesy of ABC Collection

Radio

The radio used on the I.A.R.80/81 was a Telefunken FuGVIIa radio set consisting of the S. 6b Transmitter, E. 5a Receiver, connected to each other via the VK. 5a Junction Box and powered via the UF. 4b Transformer

- Frequency range: 2500 to 3750 kHz (80 to 120 m)
- Maximum range: - 60 to 80 km from aircraft to ground,
  - 40 to 60 km from aircraft to aircraft in the air
  - 60 to 120 km from ground to aircraft when used in conjunction with a 200 Watt ground station and antenna.
- Transmitter output: 7 Watt unmodulated carrier
- Receiver sensitivity: 10 µV for 10V output voltage, 10 kΩ at 3V resistance noise

Dimensions:

- Transmitter: Height = 197 mm, Width = 339 mm; Depth = 210 mm, Weight 10.4 kg.
- Receiver: Height = 178 mm, Width = 342 mm; Depth = 245 mm, Weight 7.9 kg.
- Transformer: Weight 5.5 kg. Total weight around 32.5 kg.

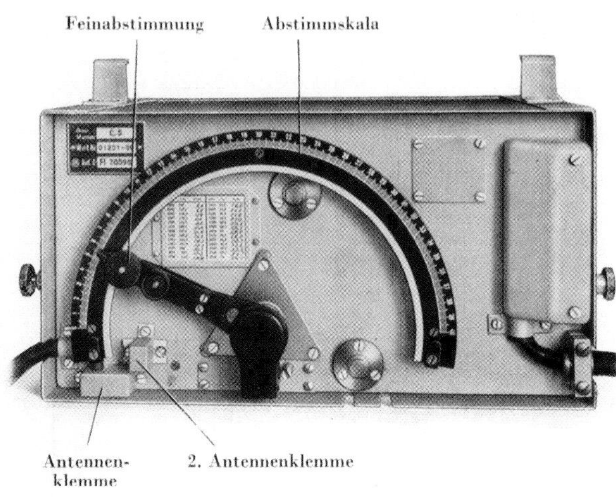

The E. 5a Receiver Courtesy of Günter Hütter

This is a photo of the FuGVIIa radio set up as a ground station. This was used by the command post in conjunction with a powerful antenna. All the main components of the radio were mounted on one single panel, inside a locker. This was identical to the radio mounted on the aircraft and only differed in the way the components were arranged. The E. 5a Receiver (top) and the S. 6b Transmitter (bottom) were mounted on the left of the panel, the Vk. 5a Junction box was mounted in the middle (top) and the U. 4b Transformer was mounted on the right of the panel (bottom). Courtesy of Günter Hütter

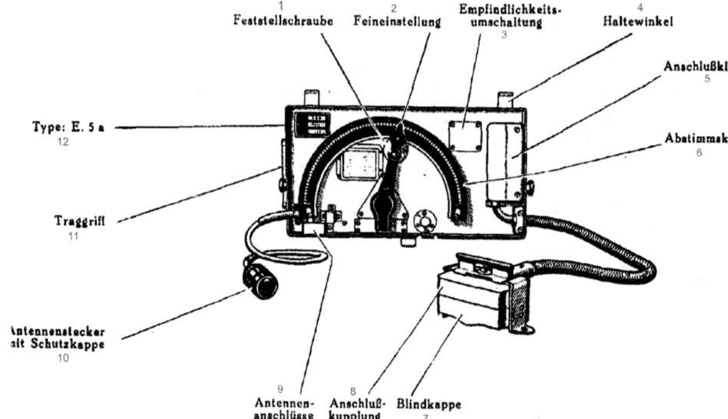

Diagram of the E. 5a Receiver. 1 = Frequency-setting knob; 2 = Fine-tuning knob; 3 = Sensitivity commuter; 4 = Hanging bracket; 5 = Connection pins; 6 = Tuning scale; 7 = Cover cap; 12 = Connection plug; 9 = Antenna connections; 10 = Antenna connection plug with protective cap; 11 = Carrying handle; 12 = E. 5a Receiver.

FN Browning 7.92 mm Machine Gun

This machine gun was made in Belgium by Fabrique Nationale d'Armes de Guerre based in Herstal. This gun was based on the Browning .30 cal machine gun design, but improved in the factory. Romania purchased a large quantity of Modéle 1932 and the Modéle 1938 machine guns. The two guns were virtually identical and differed only in the rate of fire and some minor external changes. The earlier version had a maximum rate of fire of 1200 RPM while the later version had a maximum rate of fire of 1500 RPM.

This gun was the main armament of the I.A.R.80/81 and was used in all versions.

- Gun weight — Approx. 13 kg
- Gun length of (incl. backplate) Modéle 1932 — 965 mm
- Gun length of (incl. backplate) Modéle 1938 — 985 mm
- Barrel length (muzzle to breech) — 609 mm
- Twist of grooves (right) — 254 mm
- Number of rifled grooves — 4, constant
- Ammunition — 7.92 x 57mm Mauser, fed in self-disintegrating belts
- Rate of fire (Modéle 1932) — 1200 rpm
- Rate of fire (Modéle 1938) — 1500 rpm

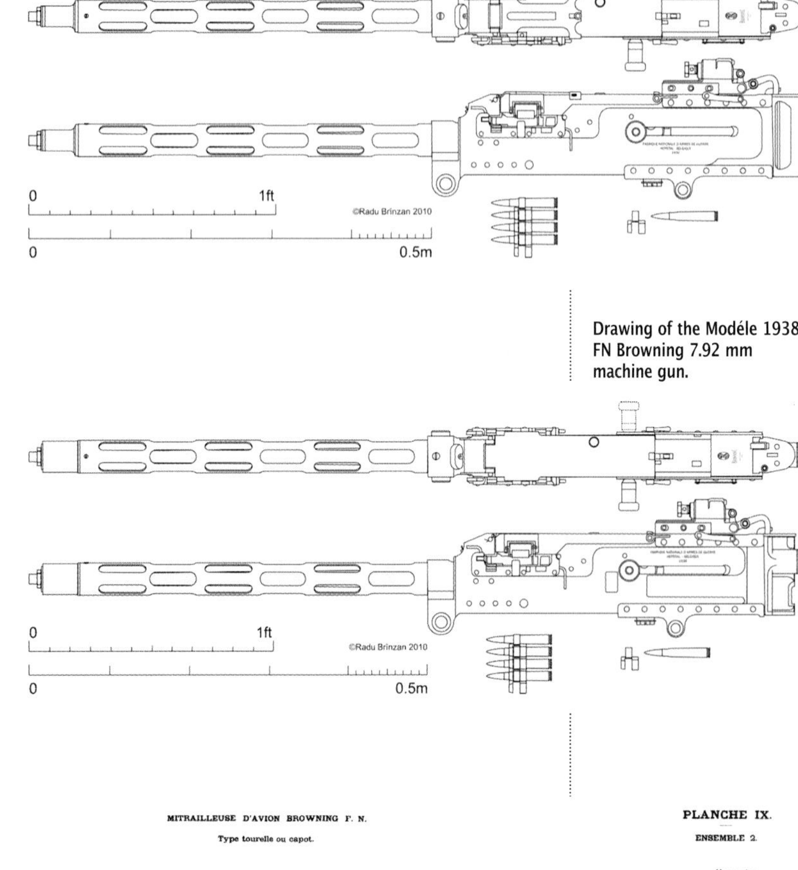

Drawing of the Modéle 1932 FN Browning 7.92 mm machine gun.

Drawing of the Modéle 1938 FN Browning 7.92 mm machine gun.

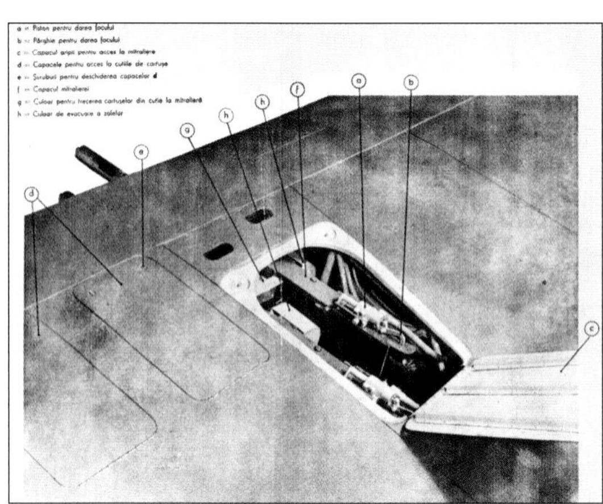

The guns of a four-gun I.A.R.80. The guns pictured here are the Modéle 1938 FN Browning 7.92 mm machine guns. Note the oblong gun cooling vents above the wing.

a = Firing piston;
b = Firing lever;
c = Machine gun servicing access panel;
d = Access panels to the ammunition magazines;
e = Opening fasteners for d panels;
f = Machine gun breech cover;
g = Supply chute bringing bullets to the gun;
h = Discharge channel for spent links.
Courtesy of ABC Collection

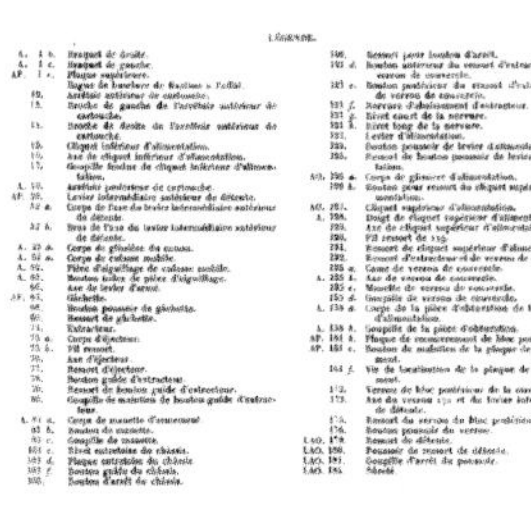

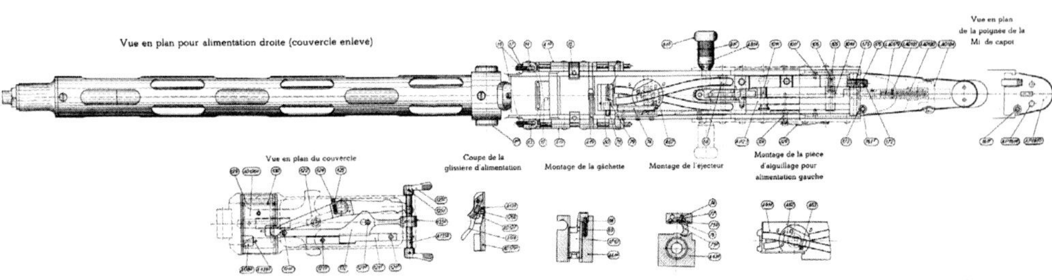

Top view of the French FN Browning 7.92 mm machine gun. This gun is fitted with a hand trigger backplate used on a flexibly-mounted gun, but the cameo drawing on the right shows the backplate used on the fixed version. Courtesy of Jean-François Legendre

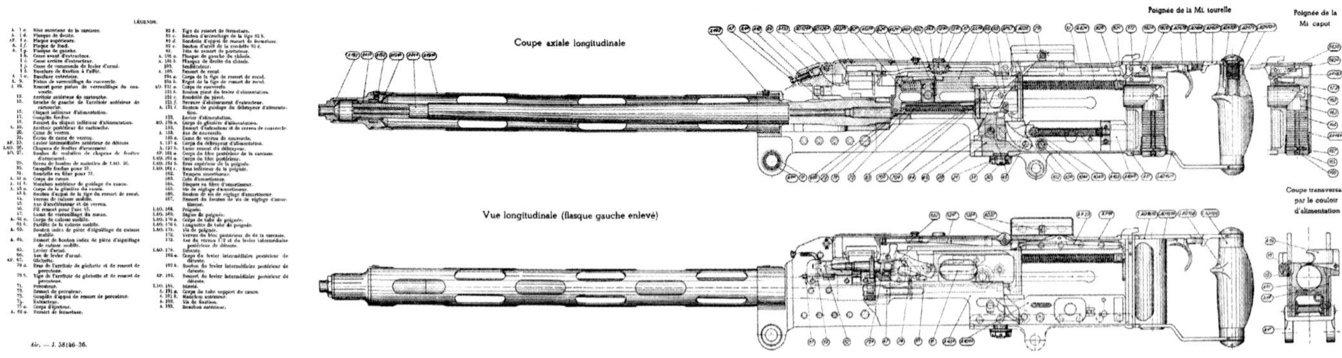

Diagram of the internal components and operation of the FN Browning 7.92 mm machine gun. The gun depicted here is the French 7.5 mm version, but with the exception of the calibre, the gun was identical to the gun used on the I.A.R.80/81. This gun is fitted with a hand trigger backplate used on a flexibly-mounted gun, but the cameo drawings on the right show the backplate used on the fixed version.
Courtesy of Jean-François Legendre

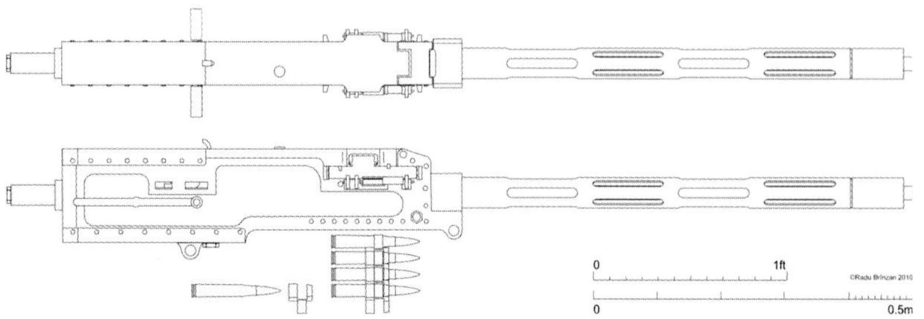

Drawing of the FN Browning 13.2 mm heavy machine gun

FN Browning 13.2 mm Heavy Machine Gun

This gun was based on the Browning M2 gun, but it was lighter and fired at a faster rate, optimised to fire the Hotchkiss 13.2x99 rounds. The gun was made in Belgium by Fabrique Nationale d'Armes de Guerre based in Herstal, under German control. This gun was used only on the I.A.R.80-B.

- Gun weight — Approx. 24 kg
- Gun length (overall) — 1450 mm
- Barrel length — 910 mm
- Twist of grooves (right) — 380 mm
- Number of grooves — 8, constant
- Ammunition — 13.2 x 99 Hotchkiss, fed in self-disintegrating belts
- Rate of fire — 600-1500 RPM

Ikaria MG-FF 20 mm Cannon

The MG FF cannon was developed in the mid-thirties by Ikaria Werke in Berlin based on the Oerlikon FFF design purchased from the Swiss company Oerlikon. The cannon was developed in a number of versions such as electrically-fired, electro-pneumatically-fired or engine-mounted. This cannon was used only on the I.A.R.80-C.

- Cannon weight — 28 kg
- Cannon length — 1,370 mm
- Barrel length — 822 mm
- Calibre — 20 mm
- Rate of fire — 520 RPM
- Muzzle velocity — 600 m/s
- Operating voltage — 24±5 Volts
- Weight of full ammunition drum — 20.3 kg

RLM scale drawing of the MG-FF cannon.
Courtesy of Erwin Wiedmer

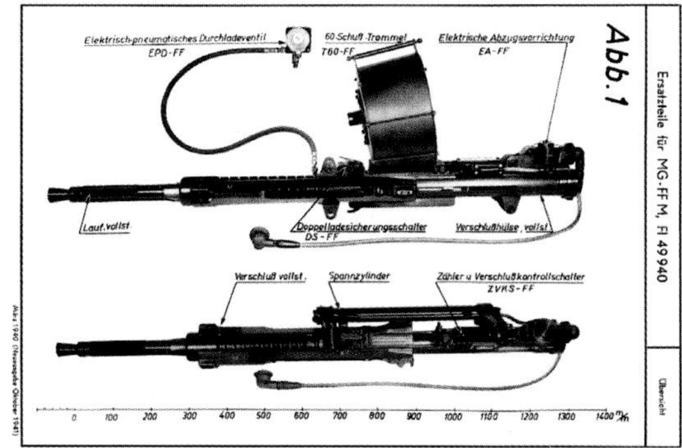

Scaled photograph showing the components of the MG-FF cannon.
Courtesy of Erwin Wiedmer

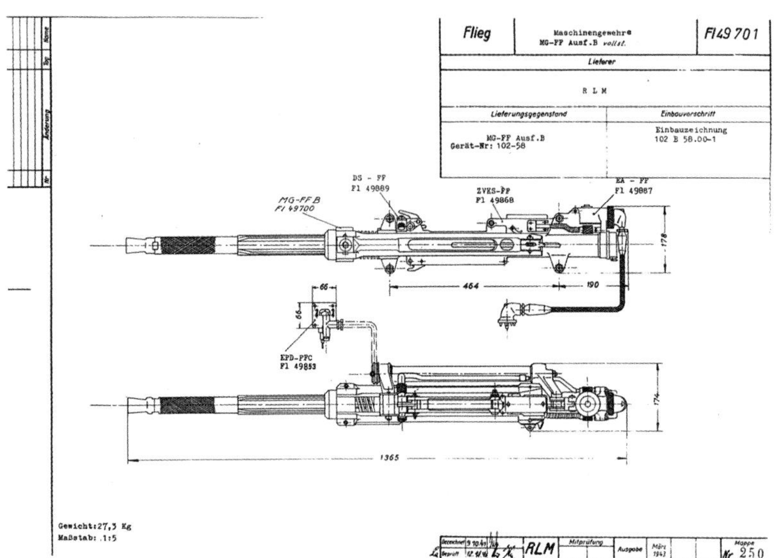

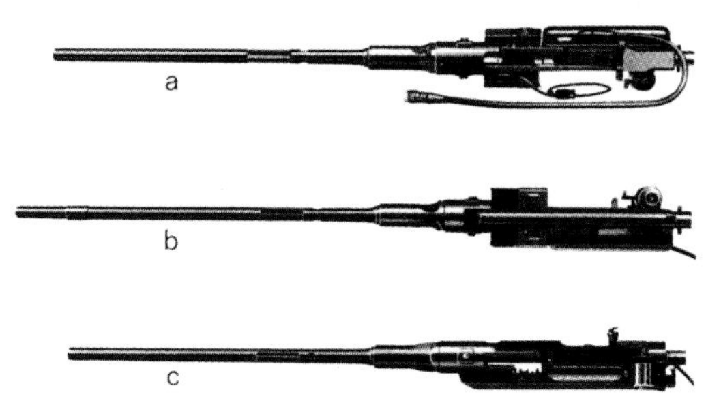

The MG151/20 cannon.

a = The cannon viewed from underneath. The large rectangular opening was the spent ammunition ejection port;

b = The cannon viewed from above. This cannon is fitted with a muzzle flash suppressor;

c = The cannon viewed from the right side (upside-down).
Courtesy of Erwin Wiedmer

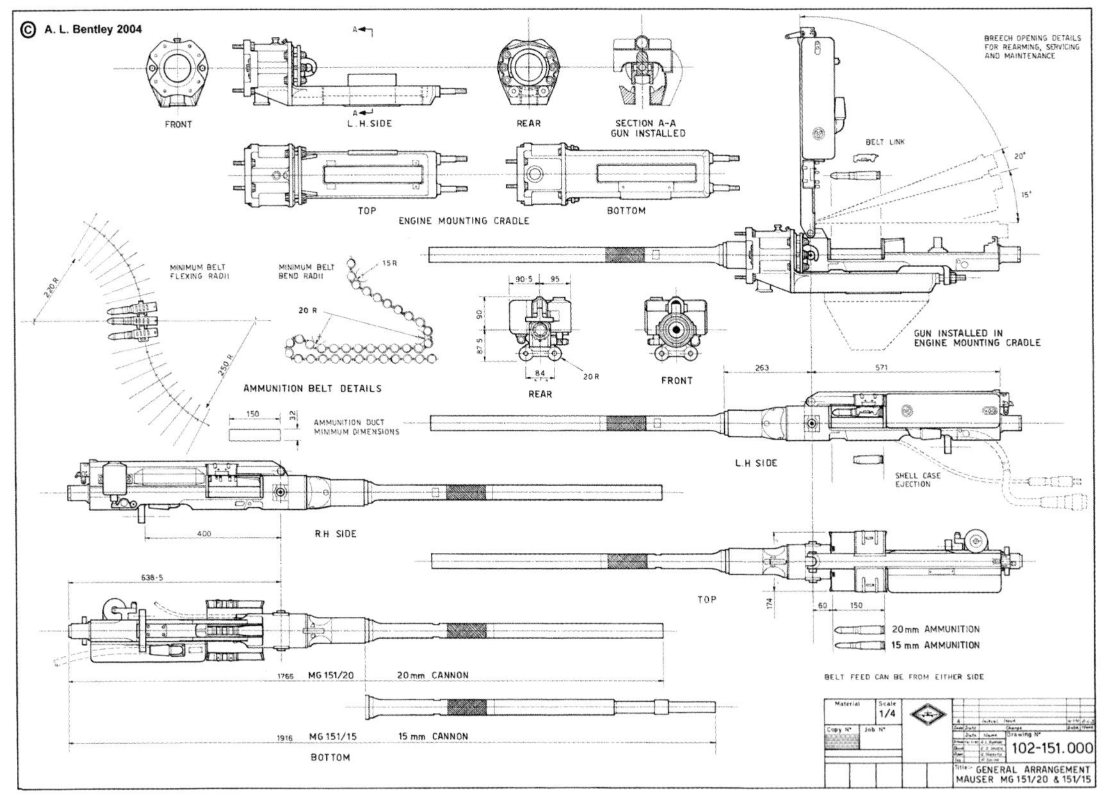

Drawing of the MG151/20 cannon by Arthur Bentley.

Mauser MG151/20 20 mm Cannon

The MG151/20 cannon was developed by Waffenfabrik Mauser AG in Oberndorf in 1941. The MG151/20 was an electrically-fired fully automatic recoil-operated cannon that could be fed from left or right with ammunition supplied in self-disintegrating belts. This cannon was fitted to the I.A.R.81-C and the modified I.A.R.80-M.

- Cannon weight — 42 kg
- Cannon length — 1,766 mm
- Barrel length — 1,104 mm
- Calibre — 20 +0.1 / -0.00 mm
- Rifling — 8 grooves, right hand twist, 1 turn in 40 cm
- Rate of fire — 750 RPM
- Muzzle velocity — 800 m/s
- Operating voltage — 24±5 Volts

A group of ground crewmen service an I.A.R.81-C. Note the interior details of the armament service panels and the open cannon breech cover. The belted rounds fed into the ammunition magazines appear to be Incendiary Tracer with Self-destructing Fuse. The technical officer working under the open fuselage panel is servicing the radio, which is slightly visible behind the tubular frame.

Courtesy of Jose Fernandez

This photo of the rocket launch tubes mounted on a Focke Wulf FW190 illustrates how these devices were mounted under the wing. It is very likely that the same arrangement was used on the I.A.R.81-C No. 429.
1 = Launch tube;
2 = Guide rail;
3 = Screw bolt;
4 = Mounting bumps;
5 = Sway braces;
6 = Mount;
7 = Bomb carrier lock;
8 = Launch cable. WGr 21 technical manual

Werfer-Granate 21 Rocket Launcher

On 15 March 1944, I.A.R. Brașov began to test the use of German WGr 21 rockets on the I.A.R.81-C. Ten launch tubes and one Bf.109-G fitted with such launch tubes were made available to I.A.R. Brașov. These weapons were fitted to I.A.R.81-C No. 429. The tubes were attached to the wing bomb carriers. The firing tests took place between 14 and 29 April 1944 in Mizil. The results of the firing tests were satisfactory, but the performance of the aircraft was affected. The rockets added an extra weight of 250 kg to the weight of the aircraft and caused a 50 km/h reduction in speed. There is no further information about this project or the use of such weapons in combat. The spin-stabilised projectile could carry a high explosive charge at a speed of 1,150 km/h (716 mph) to a maximum distance of 1,200 metres (1,300 yd). The ballistic characteristics of the projectile required that the tube be mounted at an angle of 15° to the wing. The launch tubes could be jettisoned using the same method as that used for releasing the wing-carried bombs.

BoPi Dive-Bombing Equipment

The main components of the dive-bombing equipment.
1 = Fuselage bomb carrier;
2 = Central lock;
3 = Lock release mechanism;
4 = Inert-Active device;
5 = Launcher trapeze,
6 = Pyramid;
7 = Pyramid cross brace;
8 = Lateral lock, left wing;
9 = Lateral lock, right wing;
10 = Rear bomb sway braces, left wing;
11 = Rear bomb sway braces, right wing;
12 = Forward bomb sway braces, left wing;
13 = Forward bomb sway braces, right wing;
14 = Inert-Active device, left wing;
15 = Inert-Active device, right wing;
16 = Flexible cable;
17 = Collar with stopper;
18 = Sandow cord;
21 = Electrical contact, flaps;
22 = Contact selector switch;
23 = Control box;
24 = I.A.R. - I.O.R. optical collimator.

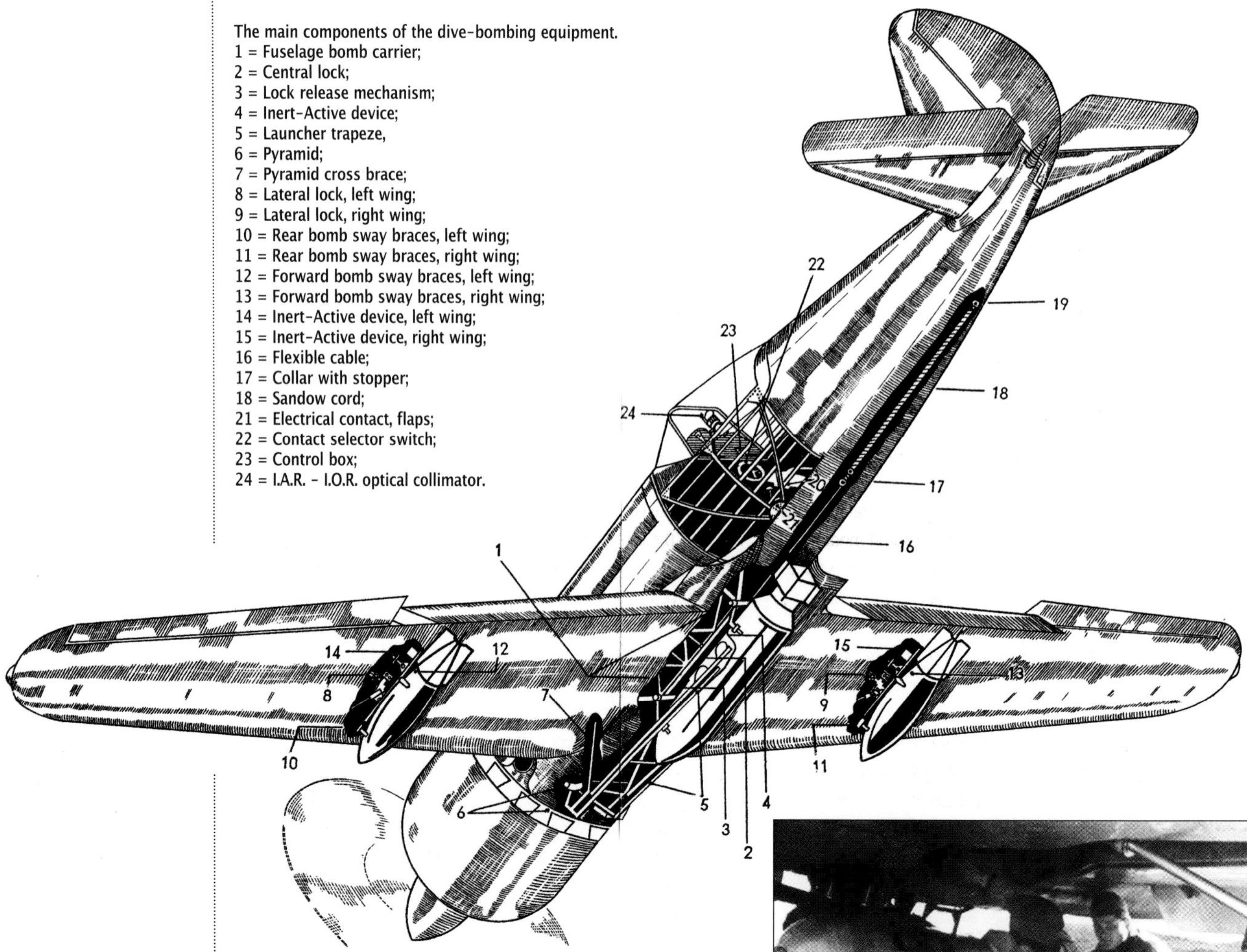

A group of armourers load an I.A.R. 225 kg bomb on the central carrier of an I.A.R.81 using a bomb lift. An I.A.R. 50 kg bomb is slung under the right wing. This photo was taken with orthochromatic film that makes the yellow engine cowl appear very dark. Courtesy of ABC Collection

After raising the bomb with the bomb lift, the armourer in the foreground secures the bomb by turning the knob in the wheel well and thus adjusting the sway braces. Courtesy of ABC Collection

An unidentified I.A.R.81 is armed with one central licence-built U.S. AN M64 500 lb bomb and two licence-built U.S. AN M30 100 lb bombs. The central bomb is decorated with a graffito, but the only readable part is 'Stalin, open your...' Courtesy of ABC Collection

The trapeze launcher was removed from this I.A.R.81-C. The stencilled test on the side next to the air supply connection is an instruction for the mechanics to raise the bomb launcher before attempting to manually start the engine [otherwise the hand crank would hit it]. The blister under the leading edge of the left wing covered one of the connections of the oil cooler underneath. 1 = Forward hinge point of the pyramid; 2 = Rear hinge point of the pyramid; 3 = Hole for the sway brace adjustment knob; 4 = Front sway braces; 5 = Bomb lock; 6 = Rear sway braces; 7 = Trapeze latch.
Courtesy of ABC Collection

Painting and Camouflage

The camouflage colours used on the I.A.R.80/81 are a veritable minefield. There are no known colour charts and the very few surviving artefacts still wearing the original colour are severely affected by the effects of the elements and the passing of time.

It is known from photographs that in the beginning, the I.A.R.80 was painted in a two-tone upper camouflage with a light blue under surface. 'Traditionally' the upper surfaces of the I.A.R.80 were depicted in drawings, illustrations and models as a relatively light tan and a light olive green, but it is hard to trace the origin of these colour tones. A quick look at the available colour photographs reveals that the colours of the two-tone camouflage were a lot darker. It is hard to pinpoint these colours with absolute precision, but in the recent years, we got tantalisingly close.

In 2009, following a conversation with Dan Cătălin Buzdugan, a Brașov-based horology-enthusiast researching Romanian aircraft clocks, I learned that he had access to the I.A.R. archives. He provided Dan Antoniu and I with the complete inventory of files held in the archives. On the last page of this amazing document, there was a list of the colours to be used on the I.A.R.80.

The colours listed were:

- Cerrux / Hexol grund gris = Cerrux / Hexol grey primer
- Cerrux / Hexol gri albastru = Cerrux / Hexol blue grey
- Hexol verde (camuflaj) = Hexol green (camouflage)
- Hexol maron (camuflaj) = Hexol brown (camouflage)
- Cerrux / Hexol albastru pt. cocarde = Cerrux / Hexol blue for cockades
- Cerrux galben pt. cocarde = Cerrux yellow for cockades
- Cerrux / Hexol rosu pt. cocarde = Cerrux / Hexol red for cockades
- Cerrux / Hexol diluant = Cerrux / Hexol thinner
- Cerrux / Hexol incolor = Cerrux / Hexol colourless
- Praf de aluminiu (Novolin) = aluminium powder (Novolin)
- Policrom (negru cu stelute) = Policrom (black with flakes)
- Cellon / Aviatin rosu de impregnare = Cellon / Aviatin red dope

These were the official names of the paints specified for the I.A.R.80 in 1941.

The Cerrux brand of paints was a cellulose-based synthetic paint manufactured by Cellon Ltd. in the United Kingdom. Cellon were involved in manufacturing aviation dopes, paints, adhesives and sealants from the beginnings of aviation and were the main supplier of aviation paints to the British Air Ministry. It is known that Cerrux paints were used, among others, on Hawker Hurricanes and Bristol Blenheims. Before the war, Romania purchased a number of Hurricanes and Blehneims directly from the UK. At that time, a certain amount of spare and replacement parts were also purchased and it is possible that the Cerrux paints suitable for the Hurricanes and Blenheims were also imported.

The Hexol brand however, seems to be rather more elusive. The brand name Hexol is used today by a Romanian branch of a Canadian manufacturer of industrial and automotive lubricants, but so far it was not possible to establish a link between this brand and the wartime paint manufacturer. Considering that in most entries on the list, Hexol and Cerrux appear to be interchangeable, it is possible that Hexol was related to Cerrux if not identical. It is also possible that Hexol was the name of a brand of paint manufactured locally under licence based on specifications that match the Cerrux paints. It is known that among the Cerrux aviation paints made by Cellon and supplied to the British Air Ministry there were two types of paint described as 'green', respectively 'Dark Green' and the countershade colour 'Light Green', and two types of paint described as 'brown', respectively 'Dark Earth' and the countershade colour 'Light Earth'. Analysis of the colour photos clearly shows that the brown and green uppersurface colours appear to be reasonably dark, which rules out the lighter countershade colours. The colours shown in these colour photos are a good match for the British Dark Green and Dark Earth camouflage. Therefore, it appears that the two-tone camouflage used on the I.A.R. may have been identical to the RAF-type camouflage applied to the uppersurfaces of the Hurricanes and Blenheims, already on strength with ARR, respectively British Dark Green and Dark Earth. Things become a lot more difficult when it comes to the underside colour. The aforementioned document specifies that it was a 'blue grey'. Among the Cerrux aviation paints manufactured by Cellon there are a number of undersurface colours that are either 'blue' or 'grey', but none that are both. The colour photographs also show that this colour was more of a light blue than a grey. The only two 'blue'

The tail plane of I.A.R.81-C No. 426 is on display in the Military Museum in Bucharest. This and the engine cowl were recovered from the crash site in Czecholsovakia and donated to the museum in the eighties. The green colour is still visible although it is in a very poor condition. Traces of red dope are still visible on the ailerons where it was used to seal and tighten the fabric. Photo by Radu Brînzan

undersurfaces on the RAF paint chart are Azure Blue and Sky Blue. The same photographs show that this colour was definitely too light to be Azure Blue. Therefore, there is a possibility that the undersurface colour used in conjunction with the two-tone uppersurface camouflage was Sky Blue, or maybe at least it was so in the beginning. Sky Blue sometimes referred to as 'duck-egg blue' is a distinctly light blue colour, which is not to be confused with the ubiquitous RAF Sky 'beige green'.

The many photographs included in this book show that the two-tone camouflage was phased out and towards the end of the war it tended to be an exception rather than a rule. Photographs indicate that beginning around 1943, the I.A.R.80/81 was mostly painted in one single colour on the uppersurfaces with a light blue fuselage underside, unlike the early 'wrap around' paint scheme. This seems to be the factory-applied paint scheme starting with No. 301 in early-1943, which is actually one third of the production. Furthermore, it is well-known that it was a common practice for ASAM to repaint the aircraft when the airframes went through their mandatory regular overhaul. This is relevant to all aircraft from the very beginning, and there is photographic evidence to show that early aircraft were painted in one single colour 'wrap around' scheme as well, and it is most likely that the early single-colour 'wrap around' green might have been the above-mentioned RAF green. However, late in the war, even the early aircraft were repainted in a new and distinctly different paint scheme with a blue tail-underside, a paint scheme most likely applied by whoever carried out the overhaul, whether it was an ASAM or the factory. This 'late' paint, applied at a time when Romania might have found it difficult to obtain RAF paints, is harder to identify. Anecdotal evidence indicates that ASAM used 'whatever they had in stores at the time' and that due to shortages of paint, they painted the aircraft in one single 'olive green' overall colour on the uppersurfaces, and 'light blue grey' on the undersurfaces. This explains the solid-green uppersurfaces with blue tail-underside paint scheme present in so many photographs. However, irrespective of the brand of paint used, these paints were undoubtedly specially-dedicated aviation paints as evidenced by their finish and durability. The brand of paints used by the ASAM remains a complete mystery and no documents related to these paints have been discovered yet, but considering that the ASAM was also used to repair Luftwaffe aircraft, it is possible that they were also supplied with German paint.

But let us return to this single uppersurface 'olive green' camouflage colour that became a standard for factory-fresh aircraft towards the end of production. 'Olive green' is not a specific shade of colour as such, and even the green colours that particularly contain the description 'olive' in their names vary in shade, such as for example the ubiquitous 'Olive Drab' that in itself is more like an 'umbrella term' covering a number of shades of green. 'Olive green' was (and continues to be) the camouflage colour of choice for many air forces because it could be easily made by mixing yellow (such as Zinc-Chromate, for example) and black (such as 'lamp black', for example). By changing the ratios, it was possible to obtain a variety of colours ranging from 'light olive green' to 'dark olive brown'. If the 'olive green' used on the I.A.R.80/81 was a workshop mix of yellow and black, then it is impossible to pinpoint the exact shade, since it may depend on the mix ratios used by the person preparing the mix.

However, another archive document found in the I.A.R. Brașov archives (Dosar. 93) reveals that at some stage in 1942 (and most likely in the subsequent two years), I.A.R. Brașov received at least two full train carriages of paint from the German manufacturer Herbig Haarhaus who were the manufacturers of a brand of lacquer-based aviation paint called Herboloid and supplied the Luftwaffe with paints based on RLM specifications. Unfortunately, this document does not point out the actual codes or names of these paints, only the quantity. So, considering that German paints were available to I.A.R. Brașov, there is a good possibility that the I.A.R.80/81 could also have been painted with German paints. But what colour?

In April 2009, Dan Antoniu and I went to the Military Museum in Bucharest where the tail-plane of I.A.R.81-C No. 426 that was shot down near Vlcnov in Czecholsovakia on 21 April 1945 while flown by Slt.Av. Gheorghe Mociorniță is on display along with a piece of the engine cowling.

The uppersurface single-colour green camouflage paint is still present and there is no evidence anywhere of any repainting or that a two-colour uppersurface camouflage has ever been applied to it. It stands to reason that this single colour was the factory-applied paint. The paint is chipped, cracked, and uneven and shows significant signs of alteration and damage. In the centre section of the wing, there is a narrow strip of relatively dark bluish-grey - this occurs in the place where the tail fin was mounted, so it is most likely a type of primer. The same grey colour, also appears as a primer inside the engine cowl, which was again covered with the same dark green. It is noteworthy that this primer is significantly darker than the light grey primer that can be seen in some photos of early unpainted I.A.R.80, which was most likely the Cerrux/Hexol Grey Primer mentioned in the itemised price list of 1941. The use of such dark primer may also be seen as a radical shift in the type of paints and method of paint application implemented late in the war. Considering the condition of the paint present on the tail plane in the Military Museum, it was impossible to identify the colour with absolute certainty, but by comparing various areas of the wing and inside of the engine cowl painted green against a Federal Standard colour-fan we reached the conclusion that the green colour was close to FS34096, FS34083 or FS33070. Understandably, the reason for the numerous matches is due to the poor condition of the paint.

This green, that may be described as an 'olive green' is similar to the late-war German RLM83 Dunkelgrun 'Dark Green' and is a little too light to be Cerrux Green or even RLM71. Knowing that German paint was supplied to I.A.R. Brașov, could this be RLM83? Considering that RLM83 was introduced by Luftwaffe in the autumn of 1944 at a time when Romania was no longer in a position to receive paint supplies from Germany and the production of the I.A.R.80/81 had already ceased, there is no

The inside of the engine cowl is covered with a dark blue-grey primer on top of which was applied the same dark olive green used on the upper surfaces of the tail plane. These colours appear to be in a better condition.
Photo by Radu Brînzan

possibility that this was RLM83. However, Kenneth Merrick mentions in his book 'Luftwaffe Camouflage and Markings 1933-1945, Volume 1' that a 'dark olive green' colour labelled RLM64 was tested by RLM before the war and manufactured in some quantity but not put into widespread use. The existing stocks of these paints were exported or sent to the Eastern Front where they were used, along with other 'pre-war colours', on FW190 and Bf109 of JG54, as evidenced by colour wartime photographs and surviving wrecks. Merrick further indicates that in the autumn of 1944, the remaining stocks of RLM64 as well as fresh batches of paint matching this formula were reintroduced under the new designation of RLM83 - a paint chart included with the book clearly shows that RLM64 and RLM83 were identical. Was RLM64 among the train-loads of paint received by I.A.R Brasov in 1942? I asked Kenneth Merrick this question and his opinion was: 'it is very possible that supplies of these colours were shipped to Romania - they were, after all, not new, merely pre-war colours that were given a new RLM number when they eventually did go back into production for the RLM. One must keep in mind that the paint companies were privately owned and did business when ever they could for pure commercial reasons. There was no restriction on these pre-war colours as the Luftwaffe had ceased using them by the start of the war, and they were not reintroduced until 1944. From the description it could be RLM 64, as the Royal Hungarian Air Force was using the same colours on its Bf 109 Fs. Manufacturing these colours was not problematic for the companies involved – simply changing the pigmentation used.' Could this explain the presence of a colour strikingly similar to RLM64(83) on the I.A.R.81-C No. 426 or is this just a coincidence? We don't know for sure, but it is a tantalising possibility.

The underside colour of Mociorniţă's I.A.R.81-C tail plane displayed in the Military Museum in Bucharest is severely oxidised and damaged, in a worse condition than the colour of the uppersurfaces. In some places where the oxidised exterior of this paint was scratched when the piece was mounted for display, the paint revealed is a very light blue grey. Because of the very poor condition of the paint and as this exposed colour was present in only small spots, in a hard to reach area without any direct illumination, it was very difficult to find an exact FS match for this colour. However, since we already considered the possibility that this aircraft might also have been painted in German colours, there is a very good possibility that the colour used on the underside was the relatively light RLM76, which subsequently became damaged by exposure and corrosion over the years.

The introduction of Soviet-style 'red star with tricolour cockade' around 1950 also saw the introduction of post-war Soviet style camouflage on the I.A.R.80/81. This was an overall single light colour. Again, there is no evidence of what this colour was, but if we are to assume that it was a paint provided by the new Soviet allies, then this colour was the post-war AGT-16 light greyish blue lacquer or A-36g greyish blue oil enamel (FS34233) used at that time for this purpose by the Soviet Air Force. AGT-16 and A-36g were effectively the same colour with the same pigment - the difference in designation indicates only the chemical composition of the carrier, respectively lacquer or enamel. However, if this overall blue-grey was again a local 'whatever was available in store' type of paint, then this can be any colour, including British Sky Blue or German RLM76 or even a 'local mix'.

One must concede that there is a lot of guessing and surmising involved in determining the true nature of the colours used on the I.A.R.80/81, but by correlating fragments of information, at least it is possible to make an informed guess about them. One can only hope that relevant documents will someday be found in the archives and these mysteries may be solved.

Cockpit Colours

Photographic evidence shows that the cockpit of the prototype was painted Aluminium Dope. It is very likely that the same type of finish was applied to early aircraft as well, but at the time of writing it was not clear how many were painted in this manner. For a brief period in 2009, an interview with Nicolae Fotescu was posted on 'YouTube' (now removed) in which the pilot recalled his battle with American fighters during the summer of 1944. In describing how he was shot down, the stated that when his plane hit the ground, the force of the impact caused the paint to flake-off from the cockpit sidewalls and when he emerged from the crash, he had to dust a significant amount of this paint from his flight suit. He described this paint as 'the same as the underside colour'. Indeed, many photographs of the cockpit of late-war aircraft undoubtedly show that at least the top part of the cockpit was painted in a light colour. Based on Mr. Fotescu's report, it is very likely that this light colour seen in the photos is the same colour as the underside light blue.

Landing Gear Colours

The wheel well was painted the same colour as the underside. In all photographs, the inside of the landing gear doors and the landing gear legs appear to be the same as the underside of the fuselage.

Propellers

The VDM propellers were painted by the German manufacturer in RLM70 Schwartzgrün with a narrow ring of unpainted natural metal at the root. Some early propellers were left unpainted (natural metal) on the front face and painted matt black on the back in order to prevent glare from the cockpit. After the war, when the aircraft were painted in Soviet-style overall grey, all areas of the aircraft were painted in the same colour, including the front and back of propeller blades. On I.A.R.80-C No.257 painted in overall Soviet-style grey, the green propeller blades were decorated with yellow tips, but this is the only such known instance.

Weathering

The I.A.R.80/81 were well maintained and were regularly repainted and cleaned. The paint finish appears to be relatively smooth and shiny and well applied. It was relatively unusual for such aircraft to feature large areas of chipped or peeled paint and when that happened, it was at the end of a long tour of duty at the frontline, especially on the Eastern Front where they flew from rough airfields and were exposed to the extreme effects of weather. Such aircraft were usually repainted as soon as they returned to the factory or ASAM for overhaul. As many photographs in this book show, most of the wear happened around the steps and handholds accessing the cockpit. On some aircraft, there is also evidence of damage caused to the top side of the forward fuselage by spilled fuel. This tended to dull the finish, which contrasted with the smoother surrounding areas. No traces of exhaust stains appear on the sides of the fuselage, even on well-worn aircraft, but there were traces of smoke on the underside of the fuselage from the three lower exhaust pipes. Some faint traces of cordite fumes may appear around the spent ammunition and link chutes as well as the machine gun barrel cooling vents. The I.A.R.80/81 was notorious for losing oil, which leaked through the oil cooler outlet on the underside and was carried by the slipstream towards the tail.

I.A.R. 80 in Colour

I.A.R.80 No. 39 of Flt.2Vt./Gr.8Vt. was photographed on Pipera airfield in the spring of 1941. The engine cowl and wingtip were painted yellow and the aircraft wore pre-war squadron markings and national insignia. During Operation Barbarossa, this aircraft was part of Esc.41Vt./Gr.8Vt. and flew on the Eastern Front from Bîrlad and Sturzeni. *Courtesy of ABC collection*

The following rare colour photographs were taken by Lt.Av. Ion Tulea in the autumn of 1942 in Brașov. At least five I.A.R.80 were photographed on the airfield of I.A.R.Brașov. The two aircraft in the middle are painted grey primer and ready to be painted, even though some panels are already covered with paint. Most likely, these were sub-assemblies that came pre-painted from the workshop, such as the tail fin, engine cowls or lateral fuselage panels. The grey primer may be Cerrux grey primer. *Courtesy of ABC collection*

I.A.R.80 No. 44, Flotila 2 Vânătoare, Grupul 8 Vânătoare, Pipera, Spring 1941

Lt.Av. Ioan Culluri stands behind the tail of the Fleet F-10G No. 2 trainer on I.A.R. Brașov airfield in the autumn of 1942. Six I.A.R.80 are parked behind him, including No. 100, No. 212, No. 87, No. 97 and No. 88. *Courtesy of ABC collection*

I.A.R.80 No. 1 and I.A.R.80-B No. 227 in September 1942. The spinners of both aircraft were painted ⅓ yellow and ⅔ green. It is interesting to note that No. 1 is still fitted with a ring and bead gunsight this late in the war. The reason why I.A.R.80 No. 1 had been sent to I.A.R. Brașov was because the left wing was damaged in September 1941 when it was hit by the propeller of I.A.R.80 No. 30 on Sturzeni airfield, on the Eastern Front while it was part of Esc.41Vt/Gr.8Vt. *Courtesy of ABC collection*

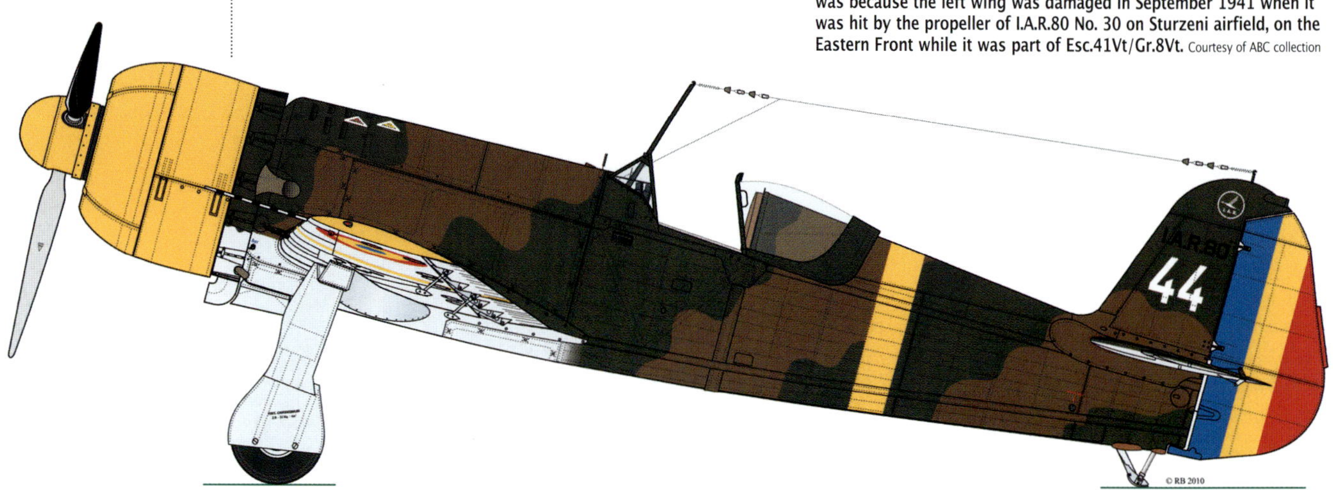

I.A.R.80 No. 1 and I.A.R.80-B No. 227 fly over the Carpathians in September 1942. When this photo was taken, these aircraft were ferried to Târgșor. I.A.R.80-B No. 227 was a factory-fresh aircraft while I.A.R.80 No.1 was returning to service after the completion of repairs. I.A.R.80 No.1 was the first production I.A.R.80 and had been used for extensive tests and trials in the last half of 1940 and the first half of 1941 by Escadrila de Experiențe in Pipera. Courtesy of ABC collection

I.A.R.80 No. 19 flown by Lt.Av. Mihail Iliescu and I.A.R.80-B No. 229 flown by Ing.Dipl. Nicolae Florescu fly over the Carpathians in September 1942.
Courtesy of ABC collection

I.A.R.80 No. 1, I.A.R.80 No. 19, I.A.R.80-B No. 227 and I.A.R.80-B No. 229 fly in formation over the Carpathians in September 1942. This photo shows to great advantage the Dark Earth and Dark Green over Sky Blue colours used on the I.A.R.80 Courtesy of ABC collection

Camouflage Schemes

This is a photo of I.A.R.80-A No. 140 of Esc.47/Gr.9Vt. based in Pipera after it crashed on 12 May 1942 with Slt.Av. Ion Galea at the controls. This photo as well as other photos published elsewhere in this book indicate that the uppersurfaces of this aircraft were painted in one single colour, most likely Dark Green. Also, it is evident that this Dark Green colour was also applied to the undersurface of the tail, in the same wraparound manner as the standard factory-applied camouflage.
Courtesy of ABC collection

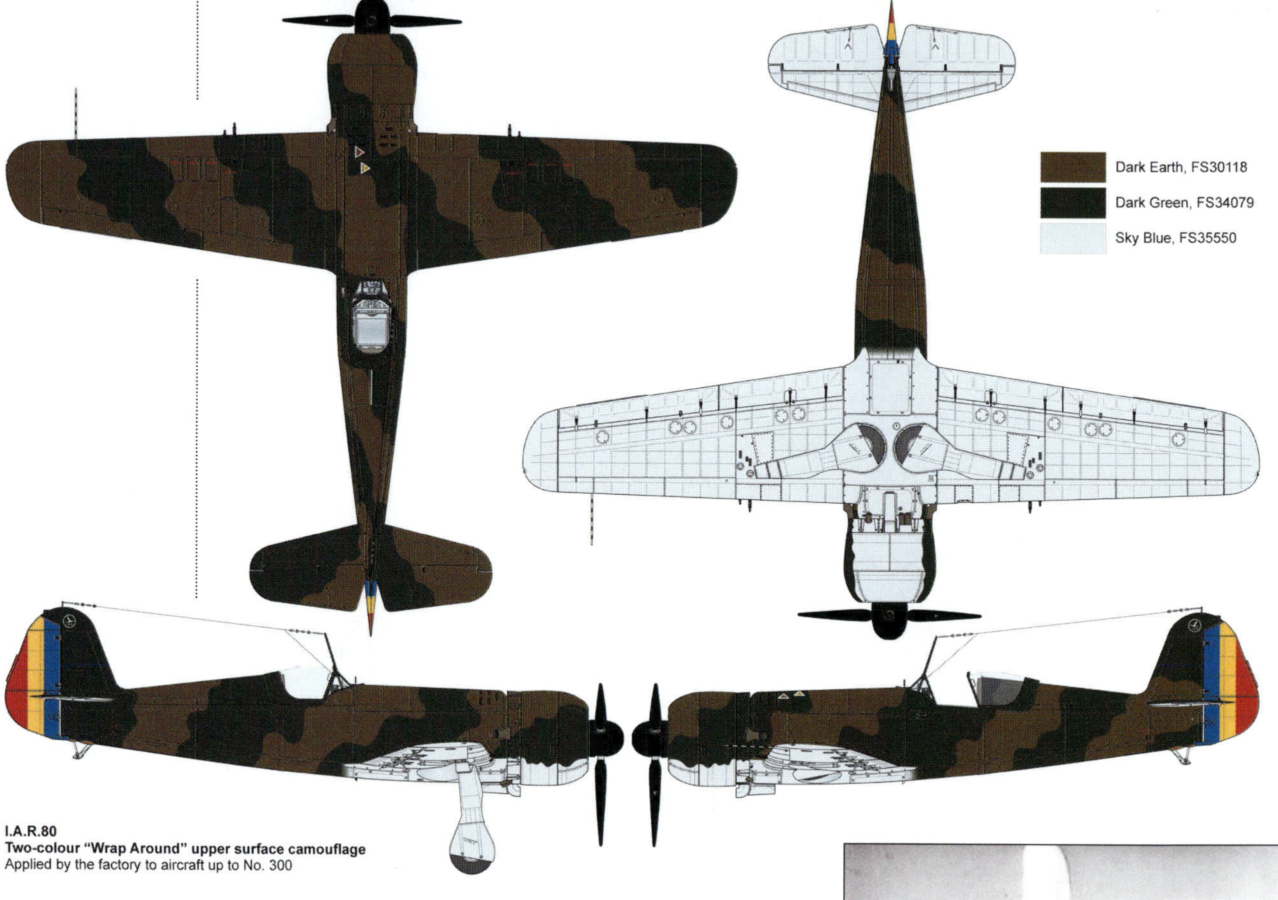

I.A.R.80
Two-colour "Wrap Around" upper surface camouflage
Applied by the factory to aircraft up to No. 300

This is a still frame taken from an 8-mm propaganda movie, but despite its poor quality, it is a very valuable illustration of the early camouflage scheme. This image shows the underside of I.A.R.80-A No. 118 of Esc.52/Gr.9Vt. based in Pipera in the summer of 1942. The Dark Earth and Dark Green upper surface camouflage wrapped is around the rear fuselage from the trailing edge of the wing to the rudder mast, while the wing, belly and tailplane undersurfaces were painted Sky Blue. This was the factory-applied camouflage scheme that was used from the beginning of production until around mid-1943. Courtesy of ABC collection

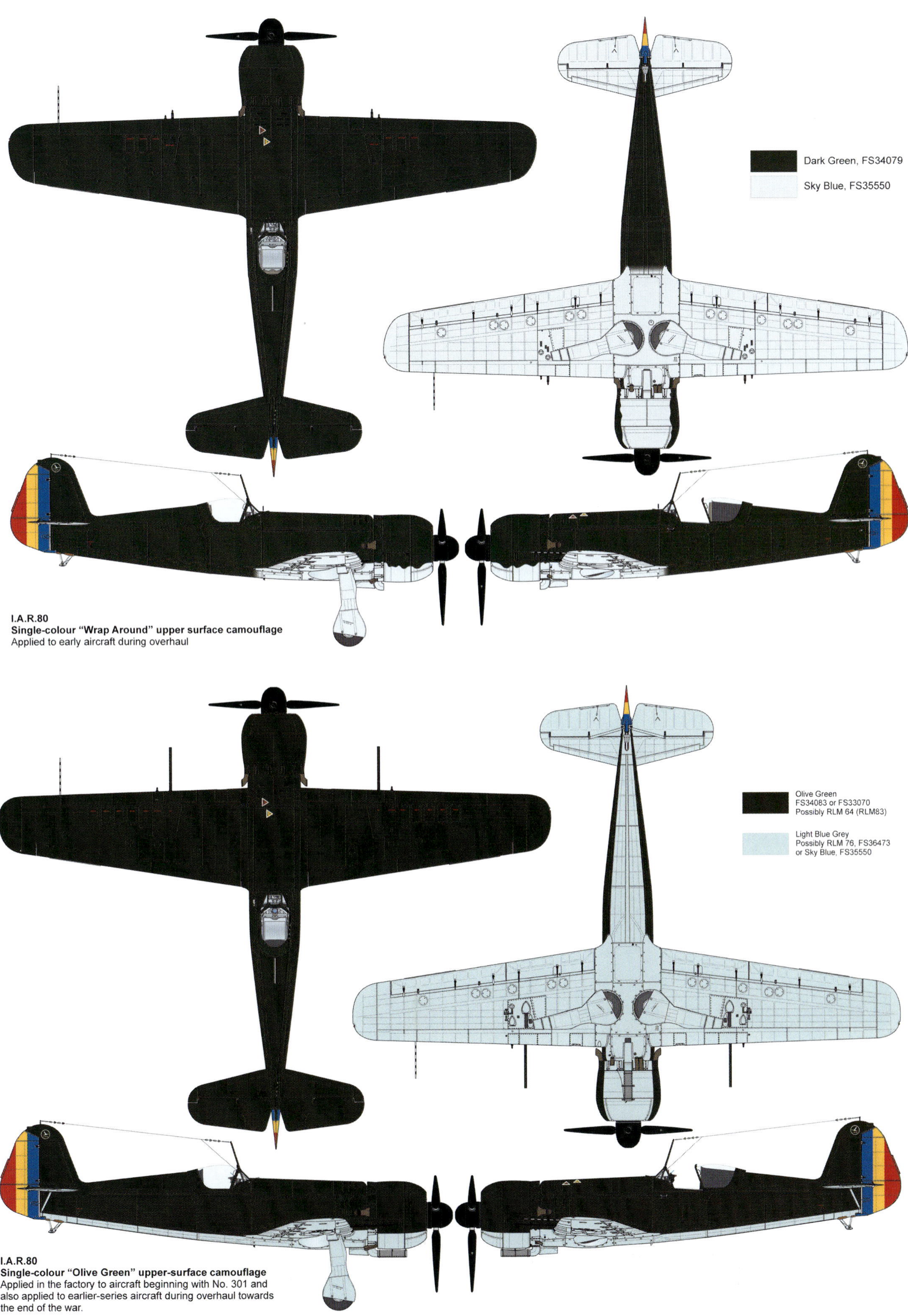

I.A.R.80
Single-colour "Wrap Around" upper surface camouflage
Applied to early aircraft during overhaul

I.A.R.80
Single-colour "Olive Green" upper-surface camouflage
Applied in the factory to aircraft beginning with No. 301 and
also applied to earlier-series aircraft during overhaul towards
the end of the war.

Markings

Cockades

The Red, Yellow, Blue cockade has always been and continues to be the national insignia of Romanian aviation, even though for some periods of time it was placed at the centre of a cross or a red star. The I.A.R.80 wore the cockade insignia until May 1941 when it was replaced by the King Michael Cross and was reintroduced in September 1944. Order No. 1312/03.09.1944 issued by SMA: 'In order to avoid regrettable and unwanted incidents that can damage the normal relations with the Soviet aviation, I order the following:

1 - Immediate measures will be taken to change the current insignia from the aircraft with the old insignia (red, yellow, blue tricolour cockade)... No aircraft shall be allowed to fly until the change of insignia... Junior State Minister for the Air, Gen.Cdt.Av. Georgescu Emil'

This underside photo of the I.A.R.80 No. 2 right wingtip was taken with orthochromatic film which makes the blue appear very faint. This type of cockade has a relatively small blue dot. ABC Collection

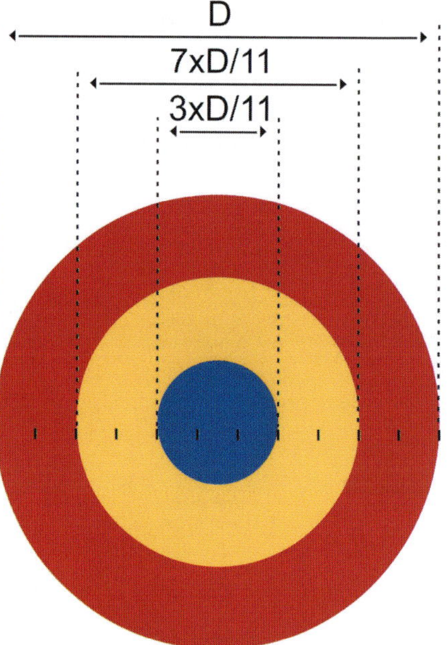

Drawing of the cockade with the smaller blue dot. Drawing by Radu Brînzan

This photo of the fuselage of I.A.R.80-A No. 121 shows a type of cockade with a relatively large blue dot. This was the type of cockade used predominantly after September 1944. ABC Collection

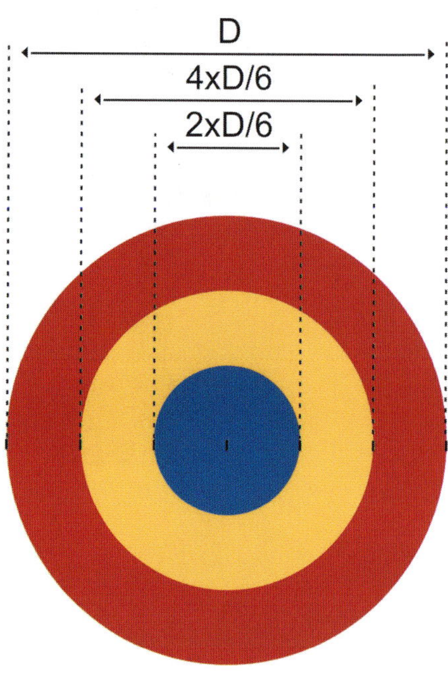

Drawing of the cockade with the larger blue dot. Drawing by Radu Brînzan

The first type of cross introduced in May 1941 had a relatively thick white outline and a very thick blue line, relatively twice as wide as the white outline. The cross had blunt tips. ABC Collection

Drawing of the first type of cross. Drawing by Radu Brînzan

Crosses

The King Michael Cross was the national insignia used on Romanian aircraft between May 1941 and September 1944.

Note No. 3798/01.05.1941 from SSA to S.S.A.

"For the purpose of telling apart the Romanian aircraft from the foreign ones, for an easier identification in the air by the German and Romanian combat personnel as well as to have an insignia more alike that used by German aviation:

1) It is my honour to report that Mr. Junior Secretary for the Air has approved and decided that the current tricolour cockades be immediately replaced from our aircraft with the "King Michael Cross" insignia, a model of which is attached to this for immediate implementation. The model is full-size and is to be applied to all aircraft except liaison, touring and school aircraft whose wing chord is less than 1mm. The model for these latter aircraft will be reduced proportionally so that they can be applied within the wing chord.

2) The application of this insignia is to be carried out immediately in the knowledge that beginning with the date of 7 May 1941, no Romanian aircraft (military or civilian) shall be allowed to fly any longer over the territory without the new insignia.

3) The application of the new insignia is to be carried out as follows:

a) On the upper surface of the upper wing and the lower surface of the bottom wing in the place of the current cockades that shall be erased. On monoplane aircraft, on both surfaces of the wing. Thus, the new insignia will be visible both from above and below on aircraft in flight.

b) On both sides of the fuselage and behind the last seat, on the highest part of the fuselage between this seat and the tail.

Note. The new insignia is not to be applied to the the rudder and elevator, which remain as they are now.

• Head of 1st Section, Cdor.Av. Gârleanu Gheorgh
• Chief of Air Staff Gen de Escadră Av. Ramiro Enescu"

Drawing of the second type of cross. Drawing by Radu Brînzan

Drawing of the third type of cross. Drawing by Radu Brînzan

The third type of cross was similar to the second type of cross but the white outline was reduced to half the previous width and the blue line was as wide as the white outline. The cross had blunt tips. ABC Collection

The forth type of cross had a thin white outline and a very thin blue cheat line. The cross had blunt tips ABC Collection

Drawing of the fourth type of cross. Drawing by Radu Brînzan

The final type of cross had a thin white outline and a very thin blue cheat line. The cross had pointed tips. This type of cross introduced around 1942 was used only on the I.A.R.80/81. ABC Collection

Drawing of the final type of cross. Drawing by Radu Brînzan

Drawing of the Romanian red star. Drawing by Radu Brînzan

Stars

The Soviet-style stars were introduced by Romania around 1950. These were also applied to the few remaining I.A.R.80 and 81, for the brief period before they were scrapped

The arms of the Romanian star were wider at the root, which differed from the similar insignia used by the neighbouring air forces. ABC Collection

Theatre markings

The I.A.R.80 wore a number of theatre markings throughout its operational life.

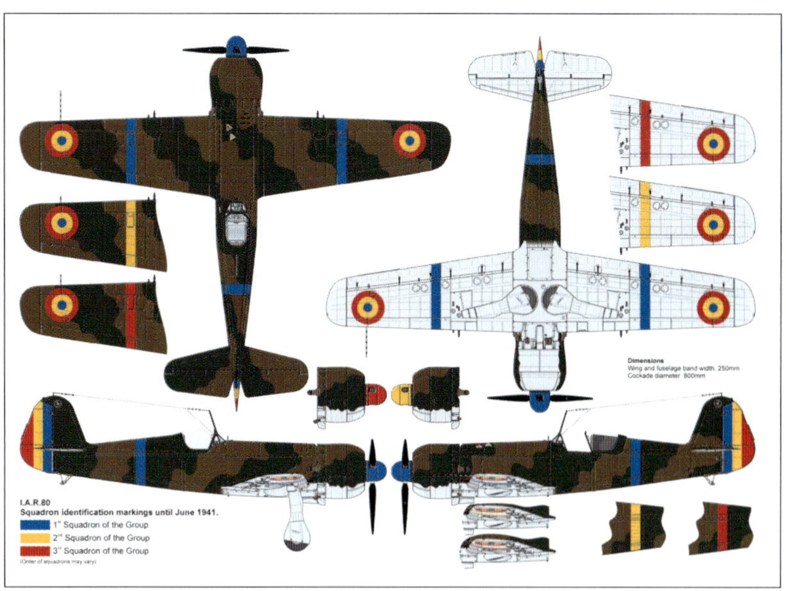

I.A.R.80
Squadron identification markings until June 1941.

- 1ˢᵗ Squadron of the Group
- 2ⁿᵈ Squadron of the Group
- 3ʳᵈ Squadron of the Group

(Order of squadrons may vary)

Dimensions
Wing and fuselage band width: 250mm
Cockade diameter: 800mm

Pre-War neutrality period: Until June 1941 (Operation Barbarossa)
The aircraft wore relatively narrow bands on the fuselage and wings denoting the squadron that they belonged to. As each group consisted of three squadrons, the colour of these squadron bands were the same as the three colours of the flag, respectively Blue, Yellow and Red. The cockade national insignia was carried only on the wings.

Pro-Axis period: After June 1941 (Operation Barbarossa)
New high-visibility recognition markings were introduced on Romanian aircraft at the time when Romania entered the war. These markings were similar to the markings used by other Axis air forces. For head-on recognition, a number of Axis Aircraft markings were painted on the propeller spinner. These included light-coloured and ⅔ dark-coloured segments or spirals.

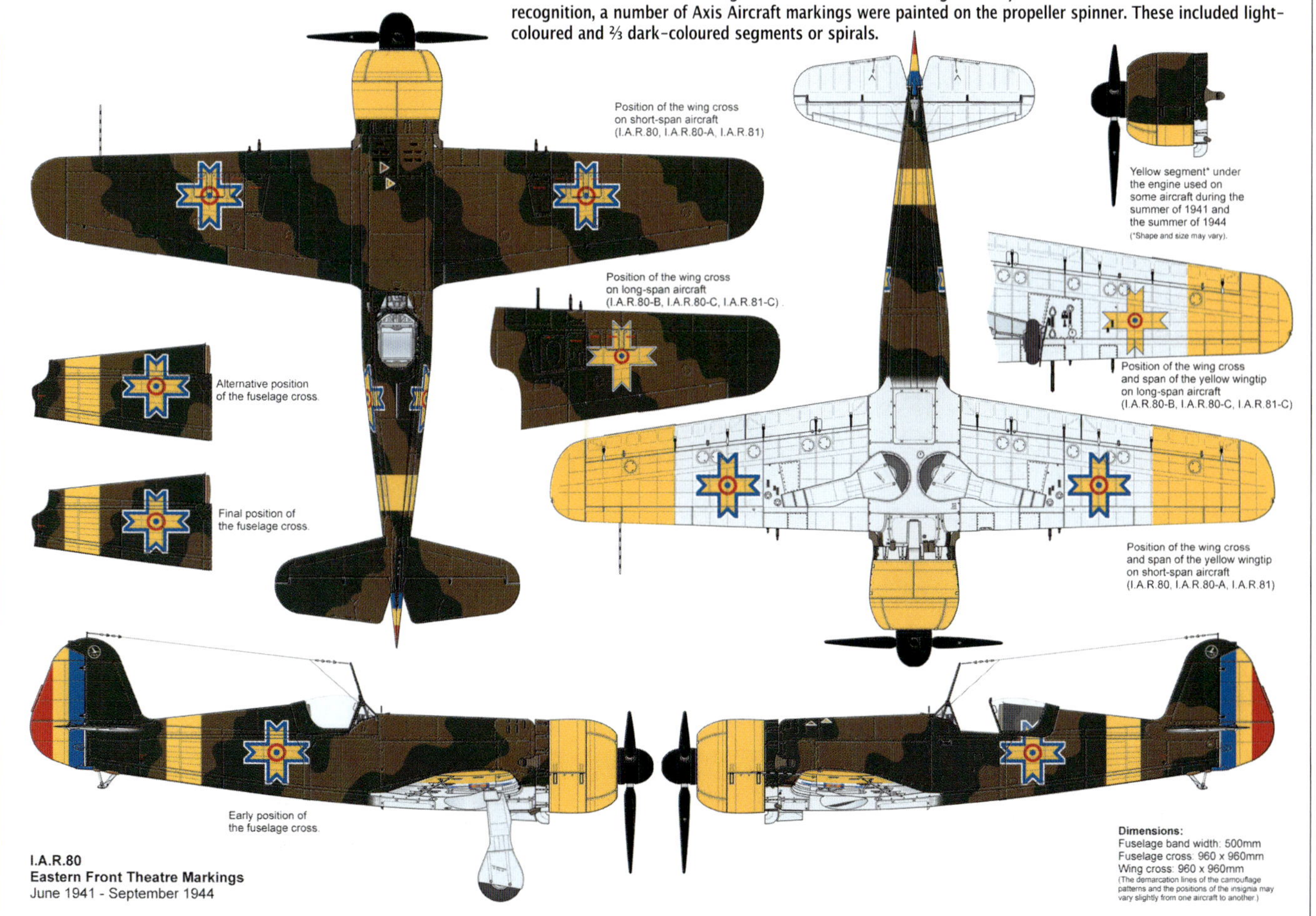

Position of the wing cross
on short-span aircraft
(I.A.R.80, I.A.R.80-A, I.A.R.81)

Position of the wing cross
on long-span aircraft
(I.A.R.80-B, I.A.R.80-C, I.A.R.81-C)

Alternative position
of the fuselage cross.

Final position of
the fuselage cross.

Yellow segment* under
the engine used on
some aircraft during the
summer of 1941 and
the summer of 1944
(*Shape and size may vary).

Position of the wing cross
and span of the yellow wingtip
on long-span aircraft
(I.A.R.80-B, I.A.R.80-C, I.A.R.81-C)

Position of the wing cross
and span of the yellow wingtip
on short-span aircraft
(I.A.R.80, I.A.R.80-A, I.A.R.81)

Early position of
the fuselage cross.

I.A.R.80
Eastern Front Theatre Markings
June 1941 - September 1944

Dimensions:
Fuselage band width: 500mm
Fuselage cross: 960 x 960mm
Wing cross: 960 x 960mm
(The demarcation lines of the camouflage
patterns and the positions of the insignia may
vary slightly from one aircraft to another.)

Pro-Allied period: After September 1944
After Romania changed sides in August 1944, there were a number of incidents caused by the markings applied to the Romanian aircraft. In one such incident, Adj. Av. Gheorghe Bucholzer flying Bf.109G-6 No. 8 was shot-down by Romanian AA fire while landing on Turnişor airfield at the beginning of the Western Campaign.

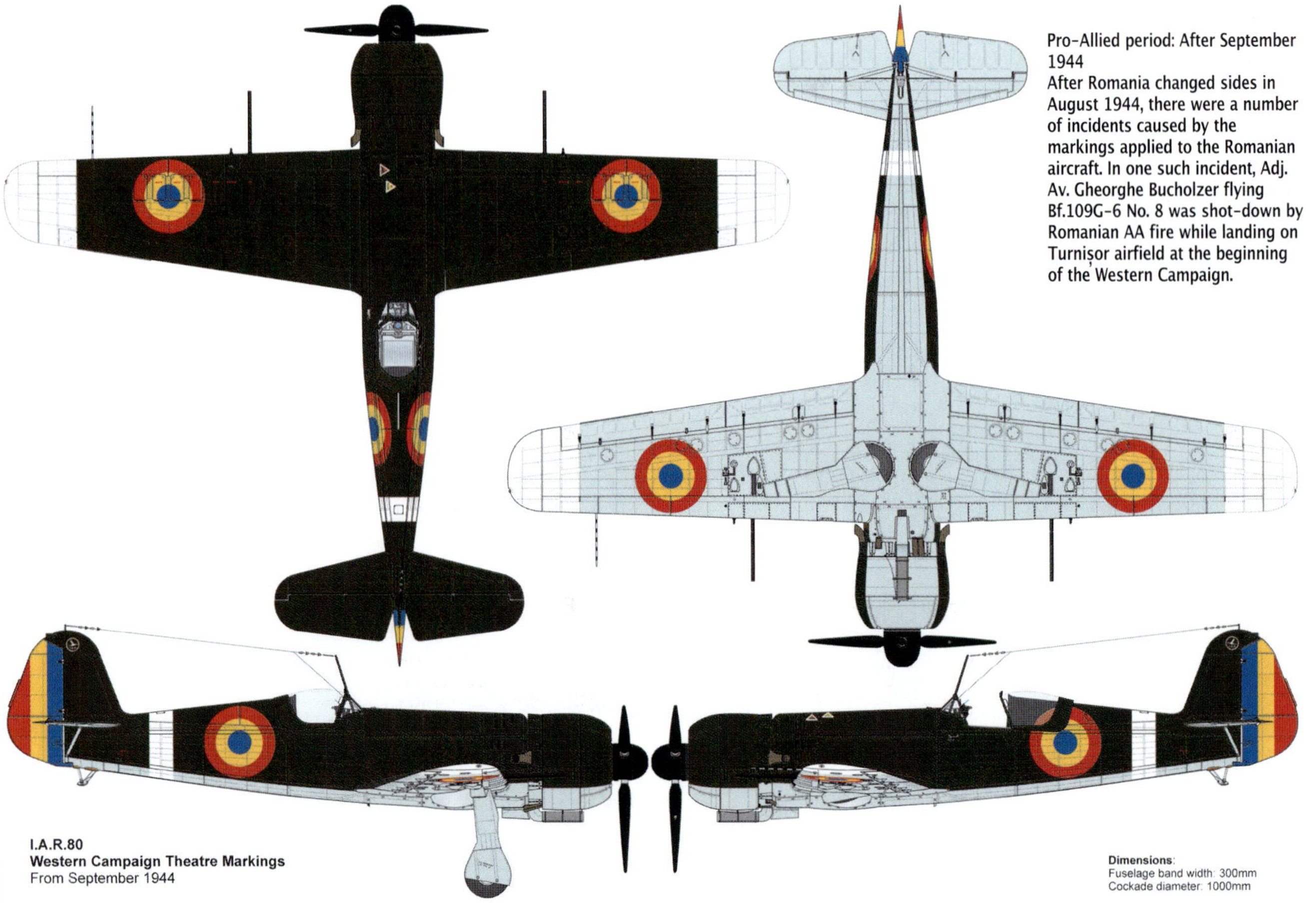

I.A.R.80
Western Campaign Theatre Markings
From September 1944

Dimensions:
Fuselage band width: 300mm
Cockade diameter: 1000mm

Pro-Soviet period: After 1950
There were no specific theatre markings associated with the Pro-Soviet period such as coloured wing tips or fuselage bands. The red-star type of insignia was introduced around 1950.

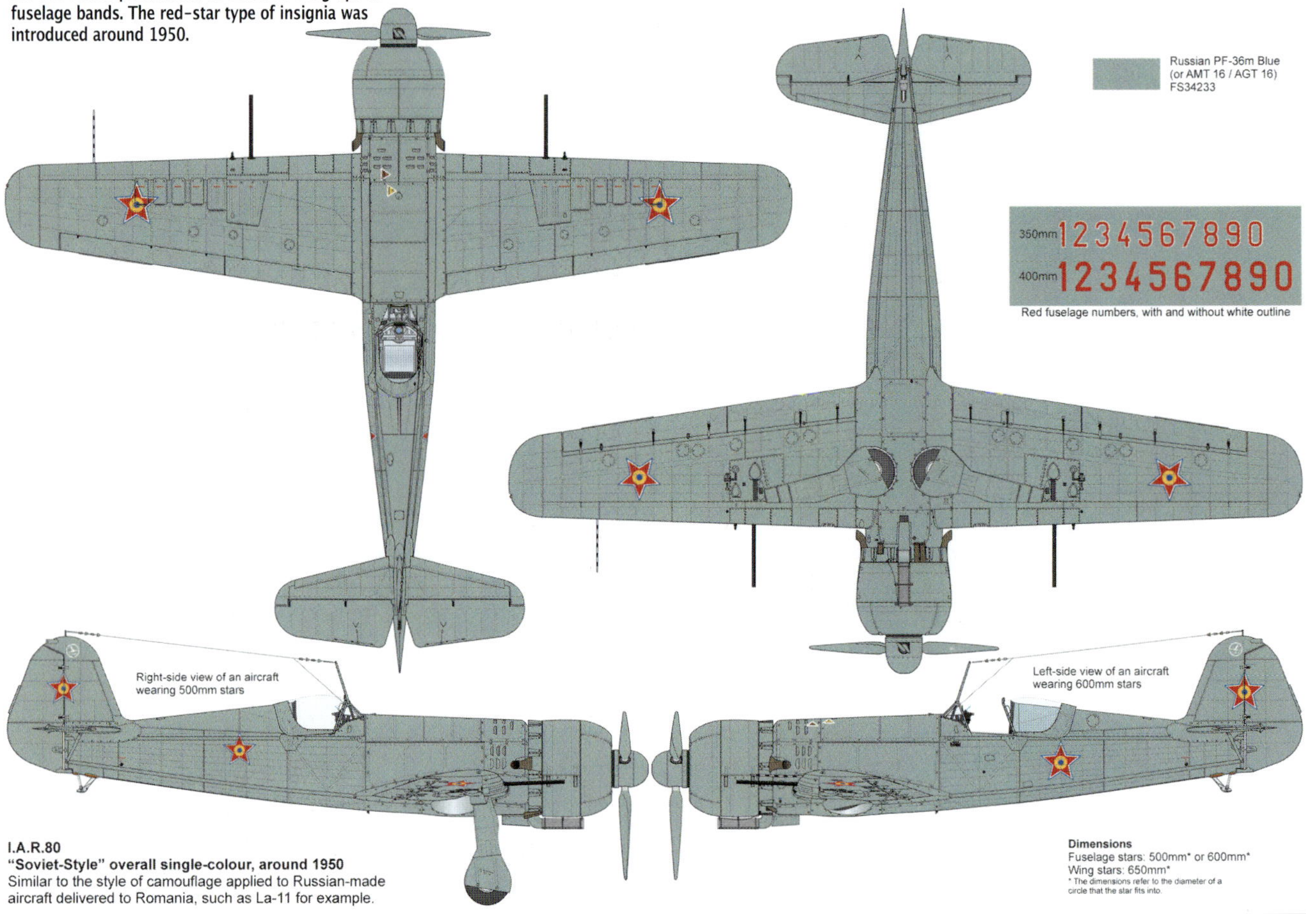

Russian PF-36m Blue
(or AMT 16 / AGT 16)
FS34233
350mm 1234567890
400mm 1234567890
Red fuselage numbers, with and without white outline
Right-side view of an aircraft wearing 500mm stars
Left-side view of an aircraft wearing 600mm stars

I.A.R.80
"Soviet-Style" overall single-colour, around 1950
Similar to the style of camouflage applied to Russian-made aircraft delivered to Romania, such as La-11 for example.

Dimensions
Fuselage stars: 500mm* or 600mm*
Wing stars: 650mm*
* The dimensions refer to the diameter of a circle that the star fits into.

Markings & Stencils

Seatbelts

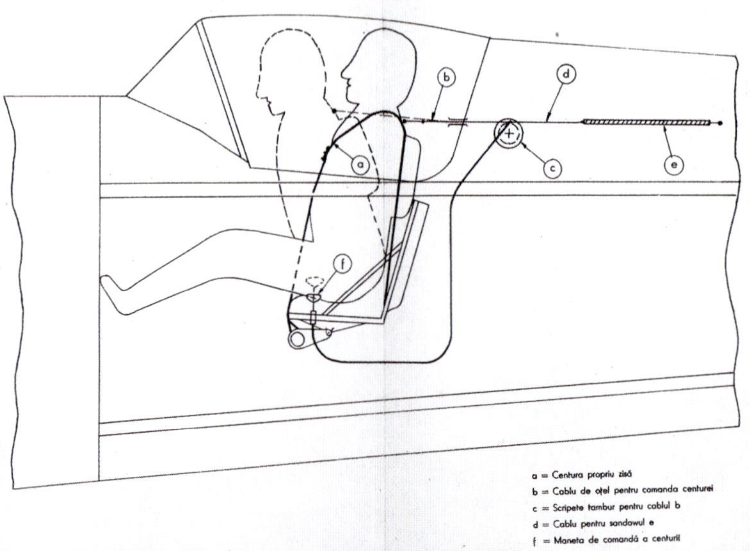

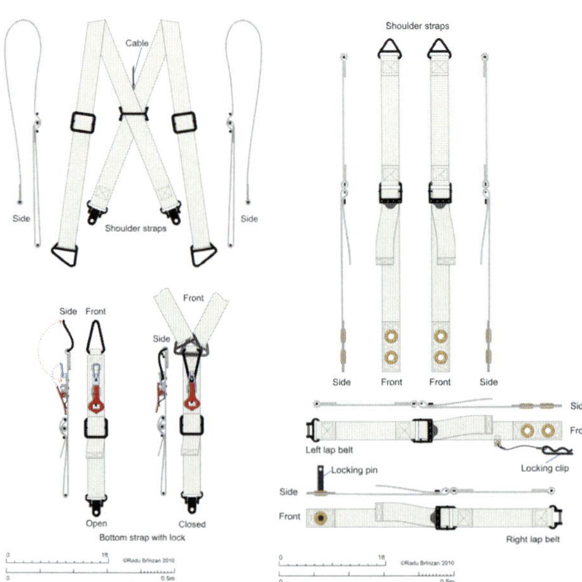

Drawing of the early type of seatbelts

Drawing of the final type of seatbelts

The initial seatbelts were an I.A.R. design. a = Seat harness straps; b = Steel cable for seatbelt control; c = Drum-pulley for the b cable; d = Cable for the bungee cord; e = bungee; f = Control lever for the seatbelt. Courtesy of ABC Collection

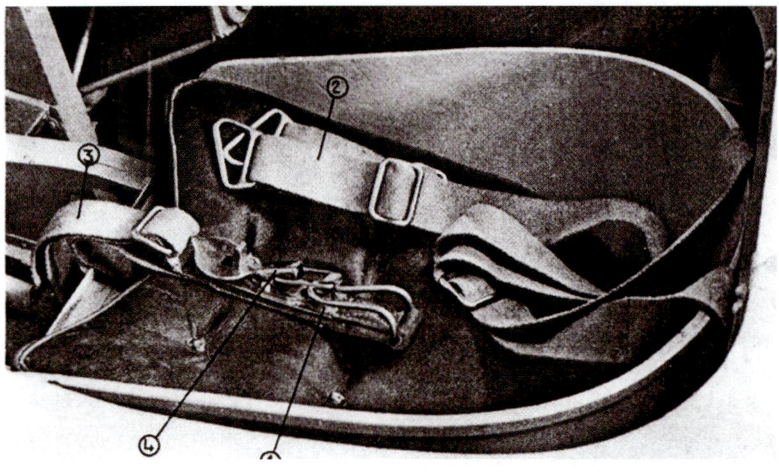

This photo shows the I.A.R. seatbelts as fitted to the I.A.R.39. 1 = Locking loop; 2 = Shoulder straps [Note the distinctive loops at the end of the straps]; 3 = Central strap [anchored to the seat pan or the floor], 4 = **Locking clasp.** Courtesy of ABC Collection

The new seat harness differed radically from the previous types and could be identified by the large fastener grommets fitted to the straps. Courtesy of ABC Collection

On 18 June 1947, I.A.R.80-C No. 254 of Flt.3 Mixtă flown by Adj. Av. Gheorghe Mazi crashed on take-off from Brașov. This aircraft was retrofitted with the final type of seat harness. A distinctive feature of the type of harness was that the shoulder straps were secured to attachment points mounted through the headrest as illustrated in this picture. Pitești Military Archives via ABC Collection